Conservation Agriculture to Soil Health and Management

Conservation Agriculture to Soil Health and Management

Nekesah T. Wafullah

Delve
PUBLISHING

www.delvepublishing.com

Conservation Agriculture to Soil Health and Management

Nekesah T. Wafullah

Delve Publishing
224 Shoreacres Road
Burlington, ON L7L 2H2
Canada
www.delvepublishing.com
Email: orders@arclereducation.com

ISBN: 978-1-77469-088-8 (Hardcover)

ABOUT THE AUTHOR

Nekesah T. Wafullah is a skilled agriculture expert with extensive knowledge in agricultural energy value addition products, agricultural business management services, project management, various forms of fertilizer, their production, sales, marketing aspects and application regimes; cross border fertilizer trade policies; youth and women empowerment and volunteerism. She is adept at project planning and management as well as creating simple solutions to complex problems. She has experience within agricultural markets in Kenya, Zambia, Malawi, Tanzania, Rwanda, Democratic Republic of Congo (DRC)-Bukavu and Lubumbashi and Uganda. She mentors high school and college students and advocates for better performance in Agricultural science. During her free time, she loves editing books, watching movies, cooking, baking, networking, reading, and dancing. Nekesah holds an M Sc. In Agricultural and Applied Economics degree from the University of Nairobi with a major in International Trade and Policy.

TABLE OF CONTENTS

LIST OF FIGURES

LIST OF TABLES

LIST OF ABBREVIATIONS

BCE	Before the Common Era
USDA	United States Department of Agriculture
CEC	Cation-Exchange Capacity
AEC	Anion Exchange Capacity
FAO	Food and Agriculture Organization
US	United States
LADA	Land Degradation Assessment in Dryland
GLASOD	Global Assessment of Soil Degradation
IFDC	International Fertilizer Development Centre
ECe	Electrical Conductivity of Extract
UNCCD	United Nations Convention on Desertification
OECD	Organization for Economic Co-operation and Development
PAHs	Polycyclic Aromatic Hydrocarbons
SOM	Soil Organic Matter
BECCS	Bioenergy With Carbon Capture and Storage
SOC	Soil Organic Carbon
USA	United States of America
CT	Computed Tomography
HRCT	High-Resolution X-Ray Computed Tomography
PAM	Polyacrylamide
NRCS	Natural Resources Conservation Service
BMPs	Best Management Practices
VFS	Vegetative Filter Strips
CREAMS	Chemicals, Runoff, and Erosion from Agricultural Management Systems
NVZs	Nitrate Susceptible Zones
EC	Electrical Conductivity
CAFOs	Concentrated Animal Feeding Operations

CRF	Controlled-Release Fertilizer
FBMPs	Fertilizer Best Management Practices
ESN	Ecologically Smart N
UDDT	Urine-Diverting Dry Toilet
OM	Organic Matter
EPA	Environmental Protection Agency

PREFACE

Soil is very important in various social and economic development. Therefore, to continue enjoying the befits of a good soil, people must understand soil aspects involving its formation classification and conservation. This book looks at how the soil is formed, its naming, characters and some important factors (physical, chemical and biological characters) and facts that makes the soil evolve. In general, over the last four decades, increased provisioning of food, fuel, and fiber has resulted in soil deterioration and the loss of numerous supporting and regulatory services given by soils. The loss of soil characteristics and regulating ecosystem services will have an influence on ecosystem provisioning services in the long run. One of the frontiers of soil science is determining the factors that influence the stability and resilience of soils in the face of disturbance. While soil deterioration occurs naturally, manmade activities have greatly accelerated it. Furthermore, soil degradation is worsening as a result of climate change and human activity. Overgrazing, deforestation, poor land management, fuelwood harvesting, and urbanization are the five major anthropogenic drivers of soil degradation worldwide, in order of increasing magnitude.

Deferent soil degradation exists. Soil erosion is a natural occurrence that has an impact on all landforms. Soil erosion in agriculture refers to the natural physical forces of water and wind, as well as forces connected with farming activities such as tillage, eroding away a field's topsoil. Erosion encompasses three separate actions: soil detachment, movement, and deposition, whether caused by water, wind, or tillage. Chemical degradation of soil refers to unfavorable changes in soil chemical behavior as a result of human involvement, resulting in a decrease in soil quality. The breakdown of soil structure, dispersion of soil particles, closure of pores, compression and increased density, consolidation, compaction, and limited root penetration, poor infiltration, waterlogging and runoff, and faster erosion are all examples of physical deterioration of soil. These processes result in desertification in arid and semiarid areas, in addition to denudation. All these degradation types affect soil health and something must be done to revert or improve soil health. Soil health refers to a soil's ability to function as a living ecosystem that supports plants, animals, and humans. Because only living things can have "health," seeing soil as a living, breathing ecosystem represents a transformation in how we think about and manage our country's soils. Soil isn't an inert growing medium; it's home to billions of bacteria, fungus, and other species who work together to form a complex symbiotic ecosystem. By cycling nutrients, absorbing, draining, and retaining rainwater and snowmelt for use during dry periods, filtering, and buffering water to remove potential pollutants, and providing habitat for the soil biological population to flourish and diversify, this ecosystem can be managed to support plants and animals. Therefore, to improve soil health, you must first recognize that it is a living system. Soil health (better soil function) is primarily an issue of providing sufficient habitat for the diverse animals that make up the soil food web. This can be accomplished by causing as little disturbance to the soil as possible, growing as many diverse species of plants as possible, keeping living plants in the soil as often as possible, and keeping the soil always covered.

Fertilizer and amendments have positive and negative effects on soils although the users tend to aim for a better soil in order to maximize productivity. Fertilizers and soil amendments are a variety of compounds that are given to soils to help plants grow more quickly. They can be natural, like bone meal, or synthetic, like 10-10-10 fertilizer. Some must be purchased, while others can be taken for free from your surroundings. There are two types of fertilizers, mineral and organic fertilizers. Mineral fertilizers include mostly nitrogen (N), phosphate, and potash, and are also known as chemical fertilizers since they are manufactured by the chemical fertilizer industry. Inorganic fertilizers main concern is its negative effect on the environment. Most organic fertilizers come from animal manures — either the feces and urine of cattle animals or the feces of poultry, either alone or in combination with bedding material (e.g., farmyard manure, chicken litter). Plant residues, municipal effluents and biosolids, and organic by-products are among the other sources. Organic manures and leguminous N fixation have been the primary sources of nutrients for farmers for thousands of years, contributing greatly to agricultural productivity and soil quality. Environmental concern on use of fertilizers have led to states coming up with some policies that Fertilizer GAPs that ensure efficient use of fertilizers.

Chapter 1

Soil

CONTENTS

INTRODUCTION

Soil is a biologically active, porous substance that has formed in the Earth's crust's uppermost layer. Soil is one of the most important substrata for life on Earth, acting as a reservoir of water and nutrients, a filtering and breakdown medium for harmful wastes, and a participant in carbon and other element cycling across the global ecosystem. It has evolved as a result of biological, climatic, geologic, and topographic influences on weathering processes. Since the advent of agriculture and forestry in the eighth millennium BCE, there has been a practical awareness of soils and their management as a result of necessity. The Industrial Revolution in the 18th and 19th centuries put increased strain on soil to generate raw materials for trade, while the advent of quantitative science provided new options for better soil management. Around the same time, systematic investigations of chemicals that increase plant growth ushered in the study of soil as a separate scientific subject. This early investigation has led to a better understanding of soils as complex, dynamic, biogeochemical systems that are essential to the life cycles of terrestrial vegetation and soil-inhabiting species, as well as humans.

THE SOIL PROFILES

Soil Horizons

Because of geologic and climatic fluctuation across time and distance, soil characteristics vary greatly. Depending on the severity and duration of weathering, episodes of soil deposition and erosion, and patterns of landscape change, even a simple attribute like soil thickness might range from a few centimeters to many meters. Despite this diversity, soils have a distinctive structural feature that separates them from other earth materials and serves as a basis for classification: a vertical succession of layers formed by the combined operations of percolating waters and living organisms. These layers are known as horizons, and the soil profile is made up of the entire vertical sequence of horizons (see the figure). Features that represent soil-forming processes characterize soil horizons. The A horizon, for example, refers to the topmost soil layer (excluding surface litter). This is a weathered layer that has accumulated humus (decomposed, dark-colored, carbon-rich materials) and microbial biomass, which has been combined with small-grained minerals to form aggregate structures.

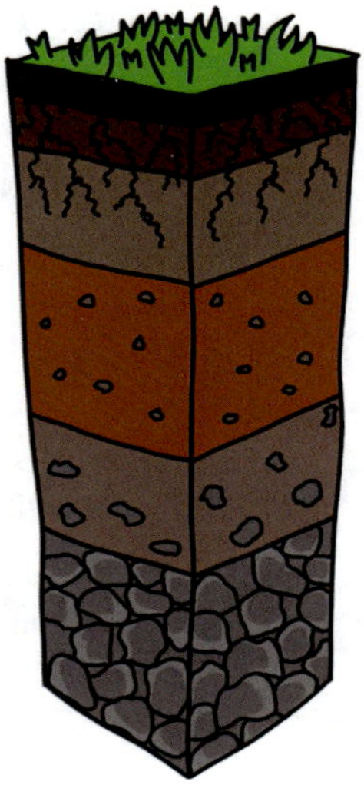

Figure 1: Soil Horizons

Source: Google Images

The B horizon is located beneath A. This layer is characterized in mature soils by a buildup of clay (tiny particles less than 0.002 mm [0.00008 inch] in diameter) that has been deposited by percolating waters or precipitated by chemical processes involving dissolved weathering products. Clay provides B horizons with a variety of structural characteristics (blocks, columns, and prisms) made from minute clay particles that can be connected in a variety of ways as the horizon develops. The C horizon, which lies beneath the A and B strata, is a zone with little or no humus accumulation or soil structure development. The C horizon is frequently made up of unconsolidated parent material that has given rise to the A and B horizons. It lacks the A and B horizons' distinguishing characteristics and can be either relatively weathered or heavily worn. Consolidated rock, which makes up the R horizon, is found at a depth below the A, B, and C horizons. There are two ways to enhance these simple letter identifiers (see the table of soil horizon letter designations). Two extra horizons are defined first. Litter and decomposed organic matter (for example, plant and animal remains) that are typically exposed on the land surface above the A horizon are designated as O horizons, whereas the layer immediately below an A horizon that has been extensively leached (that is, slowly washed of certain contents by the action of percolating water) is designated as E horizon, or zone of evaporation. High rainfall and sandy parent material enhance the development of E horizons, two characteristics that aid in broad water percolation. The solid particles that are lost due to leaching are deposited in the B horizon, which can thus be considered an illuviation zone.

Table 1: Soil horizon letter designations

Base symbols for surface horizons	
O	organic horizon containing litter and decomposed organic matter
A	mineral horizon darkened by humus accumulation
Base symbols for subsurface horizons	
E	mineral horizon lighter in colour than an A or O horizon and depleted in clay minerals
AB or EB	transitional horizon more like A or E than B
BA or BE	transitional horizon more like B than A or E
B	accumulated clay and humus below the A or E horizon
BC or CB	transitional horizon from B to C
C	unconsolidated earth material below the A or B horizon
R	consolidated rock
Suffixes added for special features of horizons	
a	highly decomposed organic matter
b	buried horizon
c	concretions or hard nodules (iron, aluminium, manganese, or titanium)
e	organic matter of intermediate decomposition
f	frozen soil
g	Gray colour with strong mottling and poor drainage
h	accumulation of organic matter
i	slightly decomposed organic matter
k	accumulation of carbonate
m	cementation or induration
n	accumulation of sodium
o	accumulation of oxides of iron and aluminium

p	plowing or other anthropogenic disturbance
q	accumulation of silica
r	weathered or soft bedrock
s	accumulation of metal oxides and organic matter
t	accumulation of clay
v	plinthite (hard iron-enriched subsoil material)
w	development of colour or structure
x	fragipan character (high-density, brittle)
y	accumulation of gypsum
z	accumulation of salts

Source: Bridges, (1993).

The solum (Latin for "floor") is the combined A, E, and B horizon sequence. The solum is the primary home for soil organisms and the true seat of soil-forming activities. (Transitional layers are identified with the two letters of the adjacent horizons because they have intermediate qualities.) The addition of lowercase suffixes to denote specific features that are relevant to soil development is the second enhancement to soil horizon terminology (also shown in the table). The most common suffixes given to B horizons are g to represent waterlogged mottling, h to denote illuvial humus accumulation, k to denote carbonate mineral precipitates, o to denote residual metal oxides, s to denote illuvial metal oxides with humus accumulation, and t to denote clay accumulation.

***O horizons or layers*:** Organic material dominates the layers. Others have never been saturated and have been saturated for lengthy periods of time or were once saturated but have been artificially drained. Un-composted or partially decomposed litter (such as leaves, needles, twigs, moss, and lichens) has been deposited on the surface of some O strata. They can be found on top of mineral or organic soils. Organic material that was deposited under saturated conditions and has degraded to various levels makes up the other O layers. The mineral component of such a material account for only a small portion of its volume and, in most cases, less than half of its weight. Some soils are fully made up of components referred to as O horizons or layers. An O layer can be found on the surface of a mineral soil or buried at any depth below the surface. Although some horizons created by the illuviation of organic material into a mineral subsoil include significant amounts of organic matter, they are not classified as O horizons.

A horizon: Mineral horizons that have formed at or under an O horizon. They indicate complete or partial obliteration of the original rock structure1 and one or both of the following:

- a deposit of humified organic materials that is tightly mixed with the mineral fraction and is not dominated by the features of the E or B horizons (described below) or
- qualities originating from disturbances such as agriculture, pasturing, or other types of disruption

It is defined as an A horizon if a surface horizon has traits of both A and E horizons, but the feature that is emphasized is a buildup of humified organic matter. The undisturbed surface horizon is less dark than the neighboring underlying horizon in some locations, such as warm, arid climates, and contains only modest amounts of organic materials. Although the mineral fraction is intact or just slightly affected by weathering, it exhibits a unique morphology from the C layer. Because it is at the surface, such a horizon is called an A horizon. Unless cultivated, recent alluvial or eolian deposits that retain fine stratification are not termed A horizons.

E horizons: Mineral horizons with a concentration of sand and silt particles as a result of the loss of silicate clay, iron, aluminum, or a combination of these elements. All or part of the original rock structure has been obliterated in these horizons. A color of higher value or lower chroma, or both, a rougher texture, or a combination of these features distinguishes an E horizon from an underlying B horizon in the same sequum. The E horizon in certain soils is the same color as the sand and silt particles, but in many soils, coatings of iron oxides or other compounds obscure the color of the primary particles. The lighter tint of an E horizon distinguishes it from the overlying A horizon. It has fewer organic materials than the A horizon in general. Although eluvial horizons that lie within or between parts of the B horizon or extend to depths greater than that of usual observation can be awarded the letter E if they are pedogenic, an E horizon is typically near the surface, below an O or A horizon and above a B horizon.

B horizons: Horizons that have formed beneath the horizons A, E, or O. They are characterized by the complete or partial obliteration of the original rock structure and exhibit one or more of the following characteristics:

- Illuvial concentrations of silicate clay, iron, aluminum, humus, carbonates, gypsum, or silica, either alone or in combination;
- Evidence of carbonates being removed or added;

- Residual concentration of oxides;
- Sesquioxide coatings that make the horizon appear lower in color value, greater in chroma, or redder in hue without illuviation of iron;
- If volume changes accompany changes in moisture content, the alteration creates silicate clay or liberates oxides, or both, and forms a granular, blocky, or prismatic structure.
- Brittleness; or
- Strong gleying.

All the various types of B horizons are or were formerly subsurface horizons. Layers of illuvial concentration of carbonates, gypsum, or silica that are the result of pedogenic processes (and may or may not be cemented) and brittle layers that show other evidence of alteration, such as prismatic structure or illuvial accumulation of clay, are included as B horizons where they are contiguous to other genetic horizons. Layers in which clay films coat rock fragments or cover finely stratified unconsolidated sediments, whether formed in place or by illuviation; layers into which carbonates have been illuviated but are not contiguous to an overlying genetic horizon; and layers with gleying but no other pedogenic changes are examples of layers that are not B horizons.

***C horizons or layers*:** Horizons or strata that are hardly impacted by pedogenic processes and lack the qualities of O, A, E, or B horizons, excluding heavily cemented and harder bedrock. Most of them are mineral layers. The material of C layers could be like or dissimilar to the material from which the solum is thought to have evolved. Even if there is no evidence of pedogenesis, the C horizon could have been altered. Sediment, saprolite, bedrock, and other geologic elements that are moderately cemented or less cemented are included in the C strata. In most cases, the excavation difficulty in these materials is low to moderate. Some soils form in extremely weathered material, and if it does not match the requirements for the A, E, or B horizons, it is identified by the letter C. Changes that are not related to the overlaying horizons are not considered pedogenic. Even if cemented, some strata with accumulations of silica, carbonates, gypsum, or more soluble salts are included in C horizons. However, a cemented layer created through pedogenic processes is referred as a B horizon.

***R layers*:** firmly anchored in indurated bedrock. The letter R designates bedrock such as granite, basalt, quartzite, limestone, and sandstone.

Excavation difficulty is frequently above average. When wet, the R layer is sufficiently cohesive to prevent hand-digging with a spade, though it can be chipped or scraped. Heavy power equipment can be used to rip some R layers. There may be fissures in the bedrock, but they are usually too few and small to allow root penetration. Clay or another material might be used to coat or cover the cracks.

W layers**:** Water This sign denotes the presence of water within or beneath the earth. If the water layer is permanently frozen, it is marked as Wf, and if it is not permanently frozen, it is designated as W. For shallow water, ice, or snow above the soil surface, the W (or Wf) indication is not used.

Transitional and Combination Horizons**:** Horizons dominated by one master horizon's attributes but with subordinate properties from another. Such transitional perspectives are denoted by two capital-letter symbols, such as AB, EB, BE, or BC. The first of these symbols implies that the transitional horizon is dominated by the attributes of the horizon so labeled. For example, an AB horizon has characteristics of both an overlaying A horizon and an underlying B horizon, but it is closer to the A horizon than the B horizon. Even if one of the master horizons to which it allegedly forms a transition is not existent, a horizon can be defined as transitional in some instances. If the attributes of a BE horizon in a truncated soil are like those of a BE horizon in a soil where the overlying E horizon has not been destroyed by erosion, it can be identified. Even if there is no underlying C horizon, a BC horizon can be identified since it is transitional to supposed parent materials. Horizons with two different portions that have the characteristics of the two types of master horizons denoted by capital letters. A virgule (/) separates the two capital letters representing such combo horizons, e.g., E/B, B/E, or B/C. The majority of one horizon component's constituent pieces are encircled by the other. The designation can be used even if there are no horizons that are identical to one or both components, as long as the individual components can be recognized in the combined horizon. The horizon with the greatest volume is the first symbol. Because single sets of horizon designators do not cover all scenarios, considerable improvisation is required. Argic Udipsamments, for example, contain lamellae separated from one another by eluvial strata. Because describing each lamella and eluvial layer as a separate horizon is often impractical, the horizons can be blended but the components specified separately. The horizon with the most lamellae and eluvial strata is called a "E and Bt" horizon. Ap-Bw-E, Bt1-E, and Bt2-C could constitute the whole horizon sequence for these soils.

***Suffix Symbols*:** To denote specific types of master horizons and layers, lowercase letters are utilized as suffixes. Many definitions of such horizons employ the term "accumulation" to suggest that they must contain more of the material in question than is assumed to have been present in the parent material. The following are the suffix symbols and their meanings:

***a-Highly decomposed organic material*:** This sign is used in conjunction with O to denote substantially degraded organic materials having a fiber content of less than 17% (by volume) after rubbing.

***b-Buried genetic horizon*:** This symbol is used in mineral soils to identify buried layers that have evolved substantial genetic traits prior to burial. Genetic horizons may or may not have formed in the overlaying material, which may be like or dissimilar to the buried soil's presumed parent material. This symbol isn't utilized in organic soils, and it's also not used to distinguish between an organic and a mineral layer.

c-Concretions or nodules: This symbol denotes a large collection of concretions or nodules. It is necessary to use cement. Iron, aluminum, manganese, or titanium are widely used as cementing agents. Silica, dolomite, calcite, or other soluble salts are not acceptable candidates.

d-Physical root restriction: non-cemented, root-restricting layers in naturally occurring or man-made sediments or materials are denoted by this symbol. Dense basal till, plow-pans, and other mechanically compacted zones are examples.

e-Organic material of intermediate decomposition: This symbol is used in conjunction with O to denote organic compounds in the intermediate stages of breakdown. After rubbing, the fiber content of these materials ranges from 17 to 40% (by volume).

f-Frozen soil or water: This symbol denotes the presence of persistent ice on a horizon or stratum. Seasonally frozen layers and dry permafrost are not represented by the symbol.

***ff-Dry permafrost*:** This symbol denotes a horizon or layer that is consistently colder than 0 degrees Celsius and lacks enough ice to be cemented by ice. This suffix is not used for horizons or strata that are warmer than 00C at some point during the year.

g-Strong gleying: This sign implies that either iron was reduced and removed during soil formation, or that it was maintained in a reduced state by saturation with stagnant water. The chroma of most of the damaged layers is 2 or less, and several have redox concentrations. The color of decreased iron or uncoated sand and silt particles from which iron has been eliminated is represented by the low chroma. The symbol g is not used for low-chroma materials that have never been wet, such as some shales or E horizons. When g is used with B, pedogenic change is implied in addition to gleying. The horizon is classified Cg if no other pedogenic change other than gleying has occurred.

h-Illuvial accumulation of organic matter: If the sesquioxide component is dominated by aluminum but only in trace amounts, this symbol is used with B to denote the accumulation of illuvial, amorphous, dispersible complexes of organic matter and sesquioxide. Sand and silt particles are coated with the organo sesquioxide substance. These coatings have solidified, filled pores, and cemented the horizon in some horizons. If the amount of sesquioxide component is large yet the color value and chroma, moist, of the horizon are 3 or less, the symbol h is also used in combination with s as "Bhs."

i-Slightly decomposed organic material: This symbol is used in conjunction with O to denote the organic compounds that are the least degraded. After rubbing, these materials have a fiber content of 40% or higher (by volume).

j-Accumulation of jarosite: Jarosite is a potassium or iron sulfate mineral that forms when pyrite is exposed to an oxidizing environment and undergoes alteration. Jarosite has a yellowish color and a chroma of 6 or greater, while it has been found with chromas as low as 3 or 4.

***jj-Evidence of cryoturbation*:** Organic soil materials occurring as bodies and broken layers within and/or between mineral soil layers, as well as irregular and broken horizon boundaries, sorted rock fragments, and organic soil materials occurring as bodies and broken layers within and/or between mineral soil layers, are all evidence of cryoturbation. Organic bodies and layers are most frequently seen at the point where the active layer meets the permafrost.

k-Accumulation of carbonates: The accumulation of alkaline-earth carbonates, most often calcium carbonate, is indicated by this symbol.

m-Cementation or induration: This symbol denotes cementation that is continuous or almost continuous. It's only employed on horizons that are more than 90% cemented, even if they're fragmented. Physically, the cemented layer restricts root growth. Adding defined letter suffixes, singly or in pairs, might indicate the main cementing agent (or the two dominating cementing agents). The horizon suffix km denotes cementation by carbonates; qm, silica cementation; sm, iron cementation; ym, gypsum cementation; kqm, lime and silica cementation; and zm, cementation by salts more soluble than gypsum cementation.

n-Accumulation of sodium: This symbol represents a buildup of sodium that can be exchanged.

o-Residual accumulation of sesquioxide's: A residual concentration of sesquioxide's is indicated by this symbol.

p-Tillage or other disturbance: This symbol denotes that the surface layer has been disturbed by mechanical means, pasturing, or other causes. Op denotes a disordered organic horizon. Even though it is a previous E, B, or C horizon, a disturbed mineral horizon is called Ap.

q-Accumulation of silica: This symbol represents a buildup of secondary silica.

r-Weathered or soft bedrock: This symbol is used in conjunction with C to denote cemented layers (moderately cemented or less cemented). Weathered igneous rock and partially cemented sandstone, siltstone, or shale are examples. The excavation difficulty ranges from easy to difficult.

s-Illuvial accumulation of sesquioxide's and organic matter: If both the organic-matter and sesquioxide components are considerable, and either the color value or chroma, moist, of the horizon is 4 or more, this symbol is used with B to indicate an accumulation of illuvial, amorphous, dispersible complexes of organic matter and sesquioxide's. If both the organic-matter and sesquioxide components are significant, and the color value and chroma, moist, are 3 or less, the symbol is also used in combination with h as "Bhs."

ss-Presence of slickensides: The presence of slickensides is indicated by this symbol. Slickensides are formed when clay minerals swell and shear failure occurs at an angle of 20 to 60 degrees above horizontal. They can indicate the presence of other vertic features such as wedge-shaped peds and surface cracks.

t-Accumulation of silicate clay: This symbol denotes a deposit of silicate clay that has formed within a horizon and then been translocated within the horizon, or that has been pushed into the horizon by illuviation, or both. Clay buildup should be visible in at least some parts of the horizon, either as coatings on rock surfaces or in pores, as lamellae, or as bridges between mineral grains.

v Plinthite: This sign denotes the existence of a reddish, iron-rich, humus-poor material that is solid or very firm while damp and hardens permanently when exposed to the atmosphere and repeated wetting and drying.

w Development of color or structure: This sign is used in conjunction with B to denote the development of color or structure, or both, with little or no illuvial material accumulation. It is not appropriate to use it to denote a transitional horizon.

x Fragipan character: This symbol denotes a layer that has been genetically modified to have a combination of firmness and brittleness, as well as a larger bulk density than the adjacent layers. Physically, some parts of the layer are root-restrictive.

y Accumulation of gypsum: This symbol denotes a gypsum deposit.

z Accumulation of salts more soluble than gypsum: This symbol denotes a concentration of soluble salts other than gypsum.

Conventions for Using Letter Suffixes

Many master horizons and layers feature one or more lowercase letter suffixes that are indicated by a single capital letter. The following rules are in effect:

- Letter suffixes should be placed immediately after the capital letter.
- More than three suffixes are rarely used.
- If more than one suffix is needed, the following letters, if used, are written first: a, d, e, h, i, r, s, t, and w. Except in the Bhs or Crt2 horizon designations, none of these letters are used in combination in a single horizon.
- If more than one suffix is needed and the horizon is not buried, the following symbols, if used, are written last: c, f, g, m, v, and x. Some examples: Btc, Bkm, and Bsv.
- If a horizon is buried, the suffix b is written last. It is used only for buried mineral soils.

Bt is a B horizon that has a considerable buildup of clay and displays evidence of color or structure development, or both (t has precedence over w, s, and h). The proper symbol for a B horizon that is gleyed or has carbonates, sodium, silica, gypsum, or salts more soluble than gypsum or residual accumulations of sesquioxide's is g, k, n, q, y, z, or o. If illuvial clay is present, the symbol t comes before the other symbol: Bto. The suffixes h, s, and w are not used with the letters g, k, n, q, y, z, or o unless they are needed for explanatory purposes.

Vertical Subdivision

Subdividing a horizon or stratum indicated by a single letter or a combination of letters is common. Arabic numbers are added to the letters of the horizon indication for this purpose. All of the letters are followed by these digits. Within a C horizon, successive layers might be labeled C1, C2, C3, and so on. The layers may be named C1-C2-Cg1-Cg2 or C-Cg1-Cg2-R if the lower section is gleyed and the higher part is not. These rules apply regardless of the subdivision's purpose. Because of the necessity to distinguish distinctions in morphological traits like as structure, color, or texture, a horizon that could be defined by a single set of letters is subdivided in many soils. These divisions are numbered sequentially using Arabic numerals, but the numbering resets to 1 if any letter of the horizon sign changes in the profile, for example, Bt1-Bt2-Btk1-Btk2 (not Bt1-Bt2 Btk3-Btk4). If the same letter combination is used in both materials, the numbering of vertical subdivisions within a horizon is not halted at a discontinuity (marked by a numerical prefix), e.g., Bs1-Bs2 2Bs3-2Bs4 (not Bs1-Bs2-2Bs1- 2Bs2). Even if distinctions in morphology are not visible in the field, thick soil layers are sometimes partitioned during sampling for laboratory studies. The Arabic digits that follow the horizon identifiers are used to identify these subdivisions. For example, Bt1, Bt2, Bt3, and Bt4 are the four strata of a Bt horizon sampled in 10-cm intervals. The set of Arabic numerals that specifies the additional sampling subdivisions follows the first numeral if the horizon has already been subdivided due to differences in morphological traits. Bt21, Bt22, and Bt23 are three layers of a Bt2 horizon sampled in 10-cm increments, for example. Each of these sampling subdivisions can have the same description, with the addition of a phrase stating that the horizon has been subdivided solely for sampling reasons.

Discontinuities

To denote discontinuities in mineral soils, Arabic numerals are used as prefixes to horizon names (preceding the letters A, E, B, C, and R). These prefixes differ from the Arabic numbers that are used as vertical subdivision suffixes. A significant change in particle-size distribution or mineralogy that indicates a difference in the material from which the horizons have formed and/or a significant difference in age, unless the difference in age is indicated by the suffix b, is a discontinuity that can be identified by a number prefix. Symbols that identify discontinuities are only used when they can make a significant contribution to comprehending the relationships between horizons. Even if genetic horizons may have formed in the contrasting layers, the stratification common to alluvium soils is not labeled as a discontinuity until particle-size distribution differs markedly from layer to layer (i.e., particle-size classes are highly contrasting). When a soil is formed entirely of a single type of material, the entire profile is regarded as material 1 and the number prefix is removed from the sign. In a profile with two or more contrasting materials, the highest material is assumed to be material 1, but the number is removed. The second layer of contrasting material, marked 2, is the first to be numbered. The underlying contrasting layers are numbered in ascending order. Even if the material of a layer below material 2 is comparable to material 1, it is marked 3 in the series; the numbers represent material changes, not material kinds. When two or more consecutive horizons occur in the same type of material, all horizon designations in that material are given the same prefix number: Ap-E-Bt1-2Bt2-2Bt3-2BC. Across the discontinuity, the suffix numbers identifying subdivisions of the Bt horizon continue in sequential order. The Arabicnumber prefix is not used if a R layer is present beneath a soil that has formed in residuum and the substance of the R layer is assessed to be like the material from which the soil has developed. However, the prefix is utilized if the R layer is expected to yield material that differs from that found in the solum, such as A-Bt-C-2R or A-Bt-2R. The sign R is given the appropriate prefix if portion of the solum has formed in residuum: Ap-Bt1-2Bt2-2Bt3-2C1-2C2-2R. A buried horizon (denoted by the letter b) has unique challenges. It's clear that it's not in the same deposit as the horizons above it. However, some buried layers have formed in lithologically similar material to the underlying deposit. To distinguish material from such a buried horizon, no prefix is necessary. However, if the material in which a buried soil horizon developed differs lithologically from the overlaying material, the discontinuity is indicated by a number prefix, and the symbol for the buried horizon is also used, for example, Ap-Bt1-Bt2-BC-C-2ABb-2Btb1-2Btb2-2C. In organic soils, there are no discernible discontinuities between distinct types of strata. If the different layers are organic, such differences are usually identified by letter-suffix designations; if the different layers are mineral, such differences are usually identified by the master symbol.

Use of the Prime Symbol

If two or more horizons of the same sort are separated by one or more horizons of a different kind in a pedon, the horizons with the same features can be represented by similar letter and number symbols. The sequence

A-EBt-E-Btx-C, for example, identifies a soil with two E horizons. The prime symbol () is inserted after the master-horizon symbol of the lower of the two horizons with identical names, e.g., A-E-Bt-E -BtxC, to underline this feature. When appropriate, the prime symbol is given to the capital-letter horizon identifier, followed by any lowercase letter symbols: B t. When the letter designations of the two levels in question are completely identical, the prime symbol is used. In the uncommon circumstance where three layers have identical letter symbols, the lowest of these layers can be represented by double prime symbols: E. The same idea applies to the classification of organic soil layers. The prime symbol is only used to distinguish two or more horizons with similar symbols, such as Oi-C-O i-C (when the soil has two identical Oi layers) or Oi-C-Oe-C (when the soil has two identical Oi layers) (when the two C layers are of the same kind).

Pedons and Polypedons

Soils are natural components of weathered environments with varying characteristics. However, for scientific purposes, it is helpful to think of soils as pedons, or unions of modules. A pedon is the smallest landscape feature that may be classified as soil. Its depth limit is the artificial line that separates soil from "not soil" (e.g., bedrock). Its lateral dimensions must be big enough to allow examination of any horizons present—in general, an area of 1 to 10 square meters (10 to 100 square feet), taking into consideration that a horizon's thickness may be varied or even discontinuous. The pedon contains one-half of the cycle when horizons are cyclic and repeat at intervals of 2 to 7 meters (7 to 23 feet). As a result, each pedon encompasses the range of horizon variability found in tiny regions. The pedon has a surface area of 1 square meter wherever the cycle is smaller than 2 meters or all horizons are continuous and of uniform thickness. Soils are found in the landscape as polypedons, which are groupings of comparable pedons with enough area to qualify as a taxonomic unit. Polypedons are bordered on the bottom by "not soil" and on the sides by pedons with different properties.

SOIL PROPERTIES AND BEHAVIOR

Every soil has properties that fall into three distinct categories:

- Physical
- Biological
- Chemical

Every producer's decision on how to manage the land and how the land responds to certain management practices is influenced by all a soil's attributes taken together. The productivity of the soil is determined by the interaction of these qualities.

Physical Characteristics

The size and arrangement of solid particles in soils, as well as how the particles affect the transport of liquids and gases through soils, are all examples of soil physical characteristics. Weathering of rocks and minerals produces mineral particles in the soil. Organic matter in the soil is the result of microbial degradation of plant and animal remnants.

Soil Texture

Sand, silt, and clay are the mineral components of soil, and their quantities define the texture of the soil. Porosity, permeability, infiltration, shrink-swell rate, water-holding capacity, and erosion susceptibility are all controlled by soil texture. The only soil in which neither sand, silt, nor clay predominates is classified loam in the USDA textural categorization triangle. While pure sand, silt, or clay can be termed soil, a loam soil with a little amount of organic material is regarded "optimal" in conventional agriculture, because fertilizers or manure are currently employed to counteract nutrient losses owing to crop yields in the long run. A loam soil's mineral composition may be 40 percent sand, 40 percent silt, and the remaining 20% clay by weight. Soil texture has an impact on soil behavior, particularly its nutrient and water retention capacity (e.g., cation exchange capacity). Sand and silt are the results of physical and chemical weathering of the parent rock; clay, on the other hand, is most typically the result of the dissolved parent rock precipitating as a secondary mineral, except for mica weathering. The specific surface area of soil particles, as well as the imbalanced ionic electric charges within them, define their function in the fertility of soil, as measured by its cation exchange capacity. Sand is the least active, with the smallest specific surface area, followed by silt, then clay. The main benefit of sand to soil is that it resists compaction and promotes soil porosity; however, this trait only applies to pure sand, not to sand combined with tiny minerals that fill in the gaps between sand grains. Silt is mineralogically like sand, but it is more chemically and physically active due to its larger specific surface area. But it is the soil's clay composition, which has a very high specific surface area and a huge number of negative charges that gives it its high water and nutrient retention capacity. Clay soils also resist wind and water erosion better than silty and sandy soils because the particles are securely bound together, and the organic content acts as a buffer. Sand is the most stable of the soil minerals; it is made up of rock fragments, usually quartz particles, with diameters ranging from 2.0 to 0.05 mm (0.0787 to 0.0020 in). The size of silt varies from 0.05 to 0.002 mm (0.001969 to 7.9105.5 in). Clay particles are 0.002 mm (7.910 5 in) in diameter and have a thickness of only 10 angstroms (1010 m), therefore optical microscopes cannot resolve them. Clay is frequently washed downhill through the soil profile in medium-textured soils (a process known as eluviation) and accumulates in the subsoil (a process called illuviation). There is no clear relationship between the size of soil mineral components and their mineralogical nature: sand and silt particles can be calcareous or siliceous, even textural clay (0.002 mm (7.9105) might be formed of very small quartz particles or multi-layered secondary minerals. Soil mineral components belonging to the same textural class may share surface area-related qualities (e.g., moisture retention), but not chemical composition-related properties (e.g., cation exchange capacity). Rock and gravel are classified as soil components larger than 2.0 mm (0.079 in) and are eliminated before establishing the percentages of the other components and the textural class of the soil, although they are included in the name. Gravelly sandy loam, for example, is a sandy loam soil that contains 20% gravel. When a soil has a significant organic component, it is referred to as organic soil rather than mineral soil. A soil is called organic if:

- Mineral fraction is 0% clay and organic matter is 20% or more
- Mineral fraction is 0% to 50% clay and organic matter is between 20% and 30%
- Mineral fraction is 50% or more clay and organic matter 30% or more.

Table 2: Influence of Soil Texture Separates on Some Properties of Soils

Property/behavior	**Sand**	**Silt**	**Clay**
Water-holding capacity	Low	Medium to high	High
Aeration	Good	Medium	Poor
Drainage rate	High	Slow to me-dium	Very slow

Soil organic matter level	Low	Medium to high	High to medium
Decomposition of organic matter	Rapid	Medium	Slow
Warm-up in spring	Rapid	Moderate	Slow
Compact ability	Low	Medium	High
Susceptibility to wind erosion	Moderate (High if fine sand)	High	Low
Susceptibility to water erosion	Low (unless fine sand)	High	Low if aggregated, otherwise high
Shrink/Swell Potential	Very Low	Low	Moderate to very high
Sealing of ponds, dams, and landfills	Poor	Poor	Good
Suitability for tillage after rain	Good	Medium	Poor
Pollutant leaching potential	High	Medium	Low (unless cracked)
Ability to store plant nutrients	Poor	Medium to High	High
Resistance to pH change	Low	Medium	High

Source: Brady Et al., (2008).

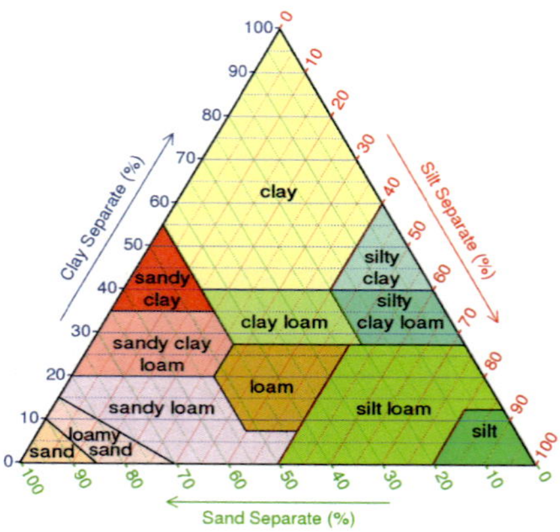

Figure 2: USDA soil textural classification triangle

Source: https://commons.wikimedia.org/wiki/File:SoilTexture_USDA.svg

Soil Structure

Aggregates are formed when the soil textural components of sand, silt, and clay clump together, and these aggregates are further associated into bigger units to form soil structures known as peds (a contraction of the word pedolith). Organic substances, iron oxides, carbonates, clays, and silica adhere to the soil textural components, and the breakage of those aggregates due to expansion-contraction caused by freezing-thawing and wetting-drying cycles, as well as the build-up of aggregates by soil animals, microbial colonies, and root tips, shape soil into distinct geometric forms. The peds develop into several types of units with different forms, sizes, and levels of development. A soil clod, on the other hand, is a mass of soil that forms as a result of mechanical disturbance of the soil, such as cultivation. Aeration, water movement, heat conduction, plant root growth, and erosion resistance are all influenced by soil structure. Water, in turn, has a significant impact on soil structure, both directly through mineral dissolution and precipitation, and indirectly through the mechanical destruction of aggregates (slaking). The texture, organic matter content, biological activity, previous soil evolution, human use, and the chemical and mineralogical conditions under which the soil developed are all revealed by the structure of the soil. While soil texture is determined by the mineral content of the soil and is an inherent quality of the soil that does not change as a result of agricultural operations, soil structure can be enhanced or damaged depending on the farming practices used and when they are used. Soil structure classes include:

- *Types*: Shape and arrangement of peds
  - Platy: The peds are flattened 1–10 mm thick one on top of the other. Found in woodland soils and lake sedimentation's A-horizon.
  - Prismatic and Columnar: Prism-like peds have a vertical size of 10–100 mm and a width of 10–100 mm. The tops of prismatic peds are flat, while the tops of columnar peds are rounded. In high sodium soil, where clay has collected, they tend to form in the B-horizon.
  - Angular and subangular: Blocky peds are imperfect cubes with sharp edges and rounded edges, measuring 5–50 mm. Where clay has collected in the B-horizon, they tend to form and signal limited water penetration.
  - Granular and Crumb: In the presence of biological material, spheroid peds of polyhedrons, 1–10 mm, are frequently

seen in the A-horizon. Crumb peds are more permeable and are said to be the best.

- *Classes*: Size of peds whose ranges are determined by the type mentioned beloe;
  - Very fine or very thin: <1 mm platy and spherical; <5 mm blocky; <10 mm prismlike.
  - Fine or thin: 1–2 mm platy, and spherical; 5–10 mm blocky; 10–20 mm prismlike.
  - Medium: 2–5 mm platy, granular; 10–20 mm blocky; 20–50 prismlike.
  - Coarse or thick: 5–10 mm platy, granular; 20–50 mm blocky; 50–100 mm prismlike.
  - Very coarse or very thick: >10 mm platy, granular; >50 mm blocky; >100 mm prismlike.
- *Grades*: Is a measurement of how well the peds have developed or cemented together, resulting in their strength and stability.
  - Weak: Weak cementation permits peds to disintegrate into sand, silt, and clay, which are the three textural elements.
  - Moderate: In undisturbed soil, peds are hardly distinguishable, but when removed, they break down into aggregates, some fragmented aggregates, and a small amount of unaggregated material. This is thought to be optimal.
  - Strong: Peds are distinct before removed from the profile and do not break apart easily.
  - Structureless: Soil is entirely cemented together in one great mass such as slabs of clay or no cementation at all such as with sand.

At the highest scale, the forces that shape a soil's structure are caused by horizontally acting swelling and shrinking, resulting in vertically oriented prismatic peds. The creation of vertisols is a good example of this mechanical process. Clayey soil will cause horizontal fissures due to its differential drying rate about the surface, reducing columns to blocky peds. Roots, rodents, worms, and freezing-thawing cycles break the peds down even more into tiny, spherical peds. Plant roots extend into voids (macropores) and eliminate water, increasing macro-porosity and decreasing microporosity, resulting in a reduction in aggregate size. Root hairs and fungal hyphae form microscopic tunnels that break up peds at the same time. Soil aggregation continues a smaller scale, when bacteria and fungi secrete sticky polysaccharides that bind soil into smaller peds. The addition of raw organic matter, which is consumed by bacteria and fungi, promotes the creation of this desired soil structure. The aggregation or dispersion of soil particles is influenced by soil chemistry at the smallest scale. Clay particles contain polyvalent cations, which cause localized negative charges on the faces of clay layers. The edges of the clay plates have a tiny positive charge, which allows them to cling to the negative charges on the faces of other clay particles or flocculate (form clumps). When monovalent ions, such as sodium, enter and displace polyvalent cations, the positive edges of the cations are weakened, while the negative surface charges are significantly strengthened. This causes the particles to push apart and deflocculates clay suspensions by leaving a negative charge on the clay faces that repels other clay. As a result, the clay disperses and settles into the voids, causing them to close. The open structure of the soil is destroyed in this manner, and the earth becomes impermeable to air and water. Near the surface, such sodic soil (also known as haline soil) tends to form columnar peds.

Soil Density

The density of soil particles ranges from 2.60 to 2.75 grams per cm3 and is usually constant for a given soil. Soil particle density is lower in soils with a lot of organic matter and greater in soils with a lot of iron oxides. Soil bulk density is calculated by dividing the dry mass of the soil by the volume of the soil, which includes air space and organic compounds. As a result, soil bulk density is always less than soil particle density, indicating that the soil is compacted. Cultivated loam has a bulk density of 1.1 to 1.4 g/cm3 (water has a bulk density of 1.0 g/cm3). Soil bulk density, unlike particle density, varies greatly within a particular soil and has a significant causal relationship with soil biological activity and management measures. Earthworms, on the other hand, have been proven to enhance or reduce soil bulk density depending on species and the size of their aggregates (feces). Due to the confounding influence of soil texture and structure, a reduced bulk density by itself does not imply suitable for plant growth. A high bulk density indicates soil compaction or a mix of soil textural classes in which small particles fill the spaces between coarser particles. As a result, there is a positive relationship between the fractal dimension of soil, which is regarded a porous medium, and its bulk density, which explains why silty clay loam has low hydraulic conductivity in the absence of a faunal structure.

Soil Porosity

Pore space is the portion of a soil's bulk volume that is not occupied by mineral or organic materials and is instead open space populated by gases or water. The total pore space in a productive, medium-textured soil is typically around 50% of the soil volume. The smallest pores (cryptopores; 0.1 m) hold water too tightly for plant roots to use; plant-available water is held in ultra-micropores, micropores, and mesopores (0.1–75 m); while macropores (>75 m) are air-filled when the soil is at field capacity. The total volume of the smallest pores is determined by soil texture; clay soils have smaller pores but more total pore space than sands, despite having a lower permeability. The bigger holes that effect soil aeration, water infiltration, and drainage are influenced by soil structure. Tillage offers the short-term benefit of temporarily increasing the number of holes of the biggest size, but the loss of soil aggregation can quickly destroy these pores. Large, continuous pores facilitate quick transmission of air, water, and dissolved nutrients through soil, while small pores store water between rainfall or irrigation episodes, affecting the ability of plants and other organisms to get water and oxygen. Pore size variation also compartmentalizes the soil pore space, allowing many microbial and faunal organisms to coexist without being in direct competition, which could explain not only the large number of species present, but also the fact that functionally redundant organisms (organisms with the same ecological niche) can coexist in the same soil.

Soil Consistency

The ability of soil to stick to itself or to other things (cohesion and adhesion, respectively) as well as its resistance to deformation and rupture is referred to as consistency. It can be used to make educated guesses about cultivation issues and foundation engineering. Air-dry, moist, and wet moisture conditions are used to determine consistency. The consistency quality in those situations is determined by the clay content. The properties of stickiness and plasticity are evaluated in the wet condition. Rubbing a soil sample in the dry condition determines its resistance to fragmentation and disintegration. Thumb and finger pressure are used to test its resistance to shearing forces in the damp state. Furthermore, cementation by components other than clay, such as calcium carbonate, silica, oxides, and salts, determines the cemented consistency; moisture content has minimal bearing on its evaluation. Because they use the apparent feel of the soil in certain stages, consistency measures are more subjective than other indicators like pH. The following phrases are used to characterize the soil consistency in three moisture conditions and one that is unaffected by moisture:

- **Consistency of Dry Soil:** loose, soft, slightly hard, hard, very hard, extremely hard
- **Consistency of Moist Soil:** loose, very friable, friable, firm, very firm, extremely firm
- **Consistency of Wet Soil:** non-sticky, slightly sticky, sticky, very sticky; non-plastic, slightly plastic, plastic, very plastic
- **Consistency of Cemented Soil:** weakly cemented, strongly cemented, indurated (requires hammer blows to break up)

The ability of soil to support structures and roadways can be estimated using soil consistency. Prior to construction, more detailed measurements of soil strength are frequently taken.

Soil Temperature

The ratio of energy absorbed to energy lost determines the temperature of the soil. Soil temperatures range from -20 to 60 degrees Celsius, with a mean annual temperature of -10 to 26 degrees Celsius depending on biomes. Seed germination, seed dormancy, plant and root growth, and nutrient availability are all influenced by soil temperature. Seasonal, monthly, and daily variations in soil temperature are significant, with fluctuations in soil temperature being significantly lower as soil depth increases. Heavy mulching (a sort of soil cover) can help to lessen surface temperature swings while also slowing the warming of the soil in the summer. Agricultural activities must frequently adapt to soil temperatures by:

- By planting at the right time, you can maximize germination and growth (also determined by photoperiod).
- Applying anhydrous ammonia to soil at temperatures below 10 °C (50 °F) maximizes its utilization.
- avoiding shallow-rooted crops from being damaged by frost heaving and thawing
- freezing waterlogged soils to prevent damage to good soil structure
- Plant phosphorus uptake is being improved.

Drying soils or using clear plastic mulches can raise soil temperatures. Organic mulches help to keep the soil from being too hot. In addition to air temperature, a variety of factors influence soil temperature, including water content, soil color, relief (slope, direction, and elevation), and soil cover (shading and insulation). The color and insulating characteristics of the ground cover have a significant impact on soil temperature. Whiter soil has a higher albedo than blacker soil, which means that whiter soils have lower soil temperatures. The energy required to raise the temperature of soil by one degree Celsius is known as the specific heat of soil. Because water has a higher heat capacity than dry soil, the specific heat of the soil increases as the water content increases. Pure water has a specific heat of 1 calorie per gram, dry soil has a specific heat of 0.2 calorie per gram, and moist soil has a specific heat of 0.2 to 1 calorie per gram (0.8 to 4.2 kJ per kilogram). Also, evaporating water requires a lot of energy (584 cal/g or 2442 kJ/kg at 25 °C) (known as the heat of vaporization). As a result, moist soil warms more slowly than dry soil; wet surface soil is normally 3 to 6 degrees Celsius cooler than dry surface soil. The pace at which heat energy travels through the soil in response to a temperature differential between two sites in the soil is referred to as soil heat flux. The heat flux density, which has both magnitude and direction, is the amount of energy that travels through soil per unit area per unit time. Seedling survival and early growth are dependent on soil temperature. The anatomical and morphological characteristics of root systems are influenced by soil temperatures. Because of the higher viscosities of water and protoplasm at low temperatures, all physical, chemical, and biological processes in soil and roots are harmed. In general, climates that do not limit white spruce survival and growth above ground are sufficient to give soil temperatures that allow white spruce root systems to thrive. White spruce grows on permafrost in some parts of the range, and while young un-lignified roots of conifers may be susceptible to freezing, the root system of containerized white spruce was unaffected by temperatures ranging from 5 to 20 °C. Temperatures between 10 and 25 degrees Celsius are ideal for tree root growth in general, and for spruce. Shoot height, shoot dry weight, stem diameter, root penetration, root volume, and root dry weight all attained a maximum at 19 °C in 2-week-old white spruce seedlings that were then grown for 6 weeks in soil at temperatures of 15 °C, 19 °C, 23 °C, 27 °C, and 31 °C. However, whereas trembling aspen and balsam poplar have demonstrated strong positive associations between soil temperature (5 °C to 25 °C) and growth, white and other spruce species have showed little or no changes in growth with increasing soil temperature. A variety of western and boreal conifers have shown to be insensitive to low soil temperatures. Soil temperatures are rising around the world as a result of current global climate warming, with conflicting perspectives on the implications on carbon capture and storage and climate change feedback loops. Permafrost thawing and its consequences for carbon destocking and ecological collapse are the most serious dangers.

Soil Color

When looking at soil, the color is frequently the first thing that comes to mind. Colors that pop and patterns that contrast are highly striking. The Red River of the South delivers material eroded from vast reddish soils such as Oklahoma's Port Silt Loam. Yellow sediment from disintegrating loess soils is carried by the Yellow River in China. Organic debris has darkened and enhanced mollisols in North America's Great Plains. Due to acidity and leaching, Podsols in boreal woods have highly different strata. The organic matter concentration, drainage conditions, and degree of oxidation all play a role in color. While soil color is easily discernible, it is ineffective in forecasting soil qualities. It can be used to differentiate horizon boundaries within a soil profile, to determine the origin of a soil's parent material, to indicate wetness and waterlogged conditions, and to measure the organic, iron oxide, and clay contents of soils qualitatively. Color is represented in the Munsell color system as 10YR3/4 Dusky Red, with 10YR denoting hue, 3 denoting values, and 4 denoting chroma. Munsell color dimensions (hue, value, and chroma) can be averaged across samples and used as quantitative characteristics, with significant relationships to soil and vegetation qualities. Soil mineralogy has the greatest influence on soil color. Various iron minerals are responsible for several soil hues. Chemical and biological weathering, particularly redox processes, are responsible for the generation and distribution of color in a soil profile. The elements combine into new and vivid compounds as the fundamental minerals in soil parent material weather. Organic waste decomposes producing black and brown humic chemicals, while manganese and sulfur can generate black mineral deposits, all of which form secondary minerals of a yellow or red tint. Within a soil, these pigments can form a variety of color patterns. Aerobic environments create uniform or gradual color changes, whereas anaerobic settings provide rapid color flow with intricate, mottled patterns and concentrated areas of color. A few words regarding soil color;

- The brown, yellow, and red colors you observe in well-aerated soils are caused by oxidized or ferric (Fe+3) iron complexes.
- When iron is converted to ferrous (Fe+2) state, it becomes mobile,

allowing it to be removed from certain soil locations. When the iron is eliminated, the remaining color is gray, or the reduced iron color is green or blue.

- Reduced iron can be re-oxidized and redeposited after aeration, sometimes in the same horizon, resulting in a mottled or variegated color pattern. The duration of the anaerobic condition can be determined by the color patterns that occur from saturation, which range from brown with a few mottles to total gray or "gleization" of the soil.
- Hydric soils have a predominate gray color with brown or yellow mottles just below the surface horizon.

Soil Resistivity

The ability of a soil to impede the conduction of an electric current is measured by its resistivity. The rate of galvanic corrosion of metallic structures in contact with the soil is influenced by the electrical resistivity of the soil. Increased moisture content or electrolyte concentration can lower resistance and increase conductivity, increasing corrosion rate. Soil resistivity levels typically range from 1 to 100000 m, with saline soils and dry soils covering crystalline rocks having the highest values.

Soil Tilth

Soil tilth is a broad term describing a soil's physical condition that is more widely used by the general population than by soil experts. Soil texture and structure both have an impact on soil tilth. Plant roots have little difficulty penetrating a soil with good tilth during their growth. It also gives plants plenty of air and water. Good tilth necessitates the existence of both macropores and micropores. Micropores store water for future plant needs, while macropores allow water to infiltrate and drain. The amount of rainfall received, as well as the pore distribution, impact whether the pores in any soil contain water or air. Capillarity is the ability of microscopic pores to retain water against gravity, which is caused by adhesive interactions between water molecules and soil particles, as well as cohesive forces between water molecules. Fine-textured soils with minimal structure and a lot of micropores may have insufficient oxygen for plant growth because small pores tend to be water-filled due to capillarity. In an air-filled pore space, oxygen diffuses quickly, while in a water-filled pore space, oxygen diffuses slowly. Tilth is a product of mechanized farming methods that damage soil structure and compact soils. The pressure imposed on soils by heavy equipment traveling over them causes compaction. Agricultural methods also tend to deplete organic matter in soils, which is necessary for structural stability.

Soil Chemical Properties

Inorganic matter in soil, organic matter in soil, colloidal properties of soil particles, soil reactions, and buffering action in acidic and basic soils are all chemical features of soils. Of course, the chemical aspect of a soil is crucial, as it pertains to the optimum balance of accessible nutrients in the soil. This is mostly controlled by the amount of organic matter and the proportion of humus in the soil; this is the farm's nutrient "storehouse." The amount of minerals present, whether dominating or not, influences the release of specific nutrients. It's necessary to fill gaps in supply, but striking the appropriate balance is even more crucial. The soil will only create nutrients if the appropriate balance is in place. Biological properties are influenced by chemical and physical qualities. Chemical and physical features that are optimal will lead to biological properties and soil functions that are optimal, such as nutrient and water cycling.

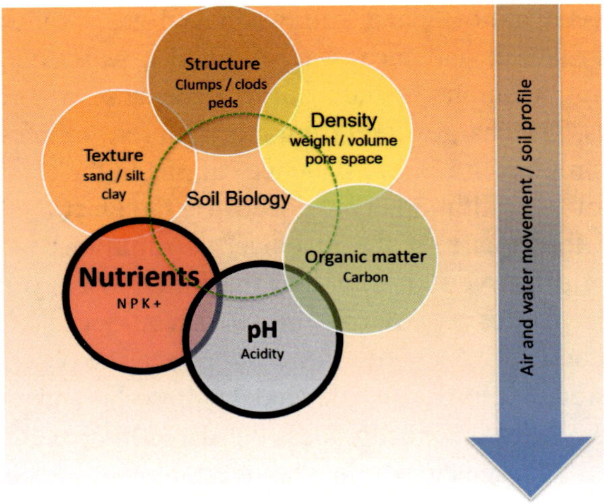

Figure 3: Interaction of soil properties

Source: Smith et al., (2018)

The chemistry of a soil affects its physical qualities and the health of its living population, as well as its ability to give available plant nutrients.

The chemistry of a soil also influences its corrosivity, stability, and ability to absorb contaminants and filter water. The chemical characteristics of soil are determined by the surface chemistry of mineral and organic colloids. A colloid is a microscopic, insoluble particle that is between 1 nanometer and 1 micrometer in size, small enough to stay suspended in a fluid medium through Brownian motion without settling. The organic colloidal particles known as humus, as well as the inorganic colloidal particles known as clays, are found in most soils. Soil's ability to store and release ions is due to colloids' large specific surface area and net electrical charges. Cation exchange occurs when negatively charged regions on colloids attract and release cations. Cation-exchange capacity (CEC) is measured in milliequivalents of positively charged ions per 100 grams of dry soil (or centimoles of positive charge per kilogram of soil; cmolc/kg). Anion exchange capacity (AEC) is determined by the ability of positively charged sites on colloids to attract and release anions in the soil.

Cation and anion exchange

By adsorbing cations of all types, both useful and detrimental, cation exchange between colloids and soil water buffers (moderates) soil pH, changes soil structure, and cleanses percolating water. Colloid particles can hold cations or anions to their surfaces because of their negative or positive charges. The charges stem from four different sources. When lower-valence cations replace for higher-valence cations in the crystal structure of clay, isomorphous substitution occurs. Because the electric charge intensity decreases with the square of the distance, substitutions in the outermost layers are more effective than those in the innermost layers. As a result, oxygen atoms have a net negative charge and can attract cations. Because the tetrahedral and octahedral structures are imperfect, the oxygen atoms at the edge of the clay are not ionically balanced. Hydroxyls can be used to replace the oxygen in the silica layers, a process known as hydroxylation. When the clay hydroxyls' hydrogens ionize into solution, they leave a negative charge on oxygen (anionic clays).

Humus hydroxyl groups' hydrogens can also be ionized into solution, leaving a negative-charged oxygen, comparable to clay. Cations bound to negatively charged colloids avoid being washed down by water and remain out of reach of plant roots, conserving soil fertility in locations where rainfall is moderate and temperatures are low. Because cations range in their strength of adsorption by the colloid and hence their ability to replace one another, there is a hierarchy in the process of cation exchange on colloids (ion exchange). If equal amounts of each are present in the soil water solution:

Al^{3+} replaces H^+ replaces Ca^{2+} replaces Mg^{2+} replaces K^+ same as NH_4^+ replaces Na^+

When a huge amount of one cation is introduced, the sheer force of its numbers may cause it to replace the others. This is referred to as the law of mass action. This is primarily what happens when cationic fertilizers are used (potash, lime). Other cations that are weaker linked to colloids are driven into solution as hydrogen ions occupy exchange sites as the soil solution gets more acidic (low pH, suggesting an abundance of H^+) (protonation). The hydrogen of hydroxyl groups may be drawn into solution at low pH, leaving charged sites on the colloid open for other cations to occupy. pH-dependent surface charges are created by the ionization of hydroxy groups on the surface of soil colloids. pH-dependent charges, unlike permanent charges created by isomorphous substitution, are variable and increase with rising pH. Freed cations can be made available to plants, but they are also susceptible to being leached from the soil, thereby depleting its fertility. Plants can excrete H^+ into the soil by synthesizing organic acids, which changes the pH of the soil near the root and pushes cations off colloids, allowing them to be available to the plant.

Cation exchange capacity (CEC)

The soil's ability to take cations from the soil water solution and sequester them to be exchanged later as the plant roots release hydrogen ions to the solution is referred to as cation exchange capacity. CEC is the quantity of exchangeable hydrogen cation (H^+) that will combine with 100 grams of dry soil, measured in milliequivalents per 100 grams (1 meq/100 g). Hydrogen ions have a single charge, and one milliequivalent of hydrogen ion equals one thousandth of a gram of hydrogen ions per 100 kilos of dry soil. Calcium converts to (40/2) x 1 milliequivalent = 20 milliequivalents of hydrogen ion per 100 grams of dry soil, or 20 meq/100 g, with an atomic weight 40 times that of hydrogen and a valence of two. The contemporary definition of CEC is centimoles of positive charge per kilogram of oven-dry soil (cmol/kg). Clay and humus colloids make up the majority of the soil's CEC, and their absence in hot, humid, rainy conditions (e.g., tropical rainforests) due to leaching and decomposition, respectively, explains tropical soils' apparent sterility. CEC is also present in live plant roots, which is linked to their surface area.

Table 3: Cation exchange capacity for soils

Soil	State	CEC meq/100 g
Charlotte fine sand	Florida	1.0
Ruston fine sandy loam	Texas	1.9
Gloucester loam	New Jersey	11.9
Grundy silt loam	Illinois	26.3
Gleason clay loam	California	31.6
Susquehanna clay loam	Alabama	34.3
Davie mucky fine sand	Florida	100.8
Sands	------	1–5
Fine sandy loams	------	5–10
Loams and silt loams	-----	5–15
Clay loams	-----	15–30
Clays	-----	over 30
Sesquioxide's	-----	0–3
Kaolinite	-----	3–15
Illite	-----	25–40
Montmorillonite	-----	60–100
Vermiculite (like illite)	-----	80–150
Humus	-----	100–300

Source: Khaing & Khaing, (2019).

Anion Exchange Capacity (AEC)

The soil's ability to extract anions (e.g., nitrate, phosphate) from the soil water solution and sequester them for later exchange as the plant roots release carbonate anions to the soil water solution is referred to as ion exchange capacity. Colloids with a low CEC are more likely to have some AEC. The maximum AEC is found in amorphous and sesquioxide clays, followed by iron oxides. Because of the generally larger rate of positively (vs negatively) charged surfaces on soil colloids, AEC levels are substantially lower than CEC levels, except for variable-charge soils. Anion exchange sites are where phosphorus is stored. Hydroxide anions (OH) in iron and aluminum hydroxide clays can be exchanged for other anions. [102] The following is the order in which anion adhesion strength is measured:

$$H_2PO_4- \text{ replaces } SO_{42}- \text{ replaces } NO_3- \text{ replaces } Cl-$$

The number of exchangeable anions per 100 g dry soil ranges from tenths to a few milliequivalents. As pH rises, more hydroxyls are formed, dislodging anions from colloids, and forcing them into solution and out of storage; thus, AEC reduces as pH rises (alkalinity).

Soil pH

The acidity or basicity (alkalinity) of a soil is measured by its pH. Soil pH is an important feature that may be utilized to conduct qualitative and quantitative analyses of soil properties. The activity of hydronium ions (H+ or, more accurately, H3O+aq) in a solution is defined as the negative logarithm (base 10) of their activity. It is generally tested in soils as a slurry of soil combined with water (or a salt solution, such as 0.01 M CaC12), and ranges from 3 to 10, with 7 being neutral. Soils with a pH below 7 are acidic, whereas those with a pH above 7 are alkaline. Ultra-acidic (pH 3.5) and extremely highly alkaline (pH > 9) soils are uncommon.

Table 4: soil pH ranges

Denomination	**pH range**
Ultra-acidic	< 3.5
Extremely acidic	3.5–4.4
Very strongly acidic	4.5–5.0
Strongly acidic	5.1–5.5
Moderately acidic	5.6–6.0
Slightly acidic	6.1–6.5
Neutral	6.6–7.3
Slightly alkaline	7.4–7.8
Moderately alkaline	7.9–8.4
Strongly alkaline	8.5–9.0
Very strongly alkaline	> 9.0

Source: Kellogg, (1993).

Because it affects so many chemical processes in soils, pH is regarded as a master variable. It has a direct impact on plant nutrient availability by regulating chemical processes and controlling the chemical forms of the various nutrients. For most plants, the ideal pH range is between 5.5 and 7.5; however, many plants have evolved to live at pH levels outside of this range. The Natural Resources Conservation Service of the United States Department of Agriculture classifies soil pH levels as follows:

Methods of determining pH include:

- Observation of the soil profile: Certain profile traits can indicate if the soil is acidic, saline, or sodic. Here are several examples:
  - Poor organic surface layer integration with the underlying mineral layer - this can suggest highly acidic soils;
  - Because podzols are highly acidic, the standard podzol horizon sequence is as follows: a pale eluvial (E) horizon sits beneath the organic top layer and overlies a dark B horizon in these soils.
  - Calcium carbonates, which are prevalent in alkaline circumstances, are indicated by the existence of a caliche layer.
  - The presence of a columnar structure can indicate the presence of a sodic state.
- Calcifuge plants (those that favor an acidic soil) include Erica, Rhododendron, and practically all other Ericaceae species, as well as numerous birch (Betula), foxglove (Digitalis), gorse (Ulex spp.), and Scots Pine (Ulex spp). (Pinus sylvestris). Ash trees (Fraxinus spp.), honeysuckle (Lonicera), Buddleja, dogwoods (Cornus spp.), lilac (Syringa), and Clematis species are all calcicole (lime-loving) plants.
- Use of a low-cost pH testing kit, which involves mixing a small sample of soil with an indicator solution that changes color depending on the acidity.
- Litmus paper is used by mixing a tiny sample of soil with distilled water and inserting a strip of litmus paper into it. If the soil is acidic, the paper turns red; if the soil is basic, the paper turns blue.
- Other pigments in fruits and vegetables change color in response to pH changes. If you add acid to blueberry juice, it turns scarlet, and if you add enough base to get a high pH, it turns indigo. Red cabbage suffers from the same problem.
- A glass or solid-state electrode is put into moistened soil or a mixture (suspension) of soil and water using a commercially available electronic pH meter; the pH is usually read on a digital display screen.

- To assess soil pH, spectrophotometric methods including the addition of an indicator dye to the soil extract have recently been developed.
- These findings compared favorably to glass electrode measurements, but they have significant advantages over glass electrode measurements, such as the absence of drift, liquid junction, and suspension effects.

For scientific study and monitoring, precise, repeatable soil pH measurements are essential. This usually requires laboratory analysis following a standard methodology; the USDA Soil Survey Field and Laboratory Methods Manual is an example of such a protocol. The mineral composition of the parent material of the soil, as well as the weathering reactions that that parent material has undergone, determine the pH of a natural soil. Soil acidification develops over time in warm, humid situations as the products of weathering are leached by water moving laterally or downwards through the soil. Soil weathering and leaching are less vigorous in dry areas, and soil pH is generally neutral or alkaline.

Base saturation percentage: There are acid-forming cations (such as hydronium, aluminum, and iron) and base-forming cations (such as calcium, magnesium, and iron) (e.g., calcium, magnesium, sodium). Base saturation refers to the percentage of negatively-charged soil colloid exchange sites (CEC) that are filled by base-forming cations. If a soil contains a CEC of 20meq and 5meq of aluminum and hydronium cations (acid-forming), the remaining spots on the colloids (20-5 = 15meq) are expected to be occupied by base-forming cations, resulting in a base saturation of 15/20 x 100% = 75%. (the compliment 25 percent is assumed acid-forming cations). Base saturation is nearly proportional to pH. (it increases with increasing pH). It can be used to figure out how much lime is needed to neutralize an acidic soil (lime requirement). The amount of lime required to neutralize a soil must consider the number of acids producing ions on colloids (exchangeable acidity), not just those present in the soil water solution (free acidity). The addition of enough lime to neutralize the soil water solution will not be enough to change the pH because the acid producing cations stored on the soil colloids will tend to restore the previous pH condition when they are driven off by the calcium of the added lime.

Buffering: The resistance of soil to changes in pH caused by the addition of acid or basic material is a measure of its buffering capacity, which increases as the CEC increases (for a certain soil type). As a result, pure sand has little buffering capacity, but soils rich in colloids (mineral or organic) contain a lot. Cation exchange and neutralization are used to buffer. However, colloids aren't the only factors that influence soil pH. Carbonates also have an important function to play. In general, many buffer systems take precedence over one other depending on pH levels, ranging from the calcium carbonate buffer range to the iron buffer range. When a modest amount of highly basic aqueous ammonia is added to a soil, the ammonium displaces hydronium ions from the colloids, leaving water and colloidally fixed ammonium as the product, with little long-term change in soil pH. The addition of a little amount of lime, $Ca\,(OH)_2$, to the soil colloids will displace hydronium ions, producing calcium fixation to colloids and the evolution of CO_2 and water, with little lasting change in soil pH. An increase in a particular cation in the soil water solution causes that cation to be fixed to colloids (buffered), while a drop in that cation's solution causes it to be removed from the colloid and transferred into solution (buffered). The degree of buffering is frequently linked to the soil's CEC; the higher the CEC, the greater the soil's buffering ability.

Soil Biological Properties

The soil is home to a diverse range of life forms, including viruses, earthworms, and huge burrowing mammals like gophers and ground squirrels. Microorganisms in the soil are referred to as "soil microflora" (though not all are plants in the literal meaning of the word), while larger creatures are referred to as "macrofauna." Soil animals, particularly earthworms and some insects, have a positive impact on the soil through their burrowing and feeding activities, which increase aeration and drainage by modifying the soil structure. In general, they have a minor impact on soil chemical characteristics, but their actions indirectly boost microbial activity by creating a more conducive soil habitat.

Soil Microorganisms

Soil microorganisms are abundant and diverse, with a wide range of shapes and functions. Bacteria (including actinomycetes), fungus, algae (including cyanobacteria), and protozoa are the most common microbiological groups in soil. Huge numbers of soil microorganisms can occupy a very small volume due to their extremely small cell size (one to several micrometers), hence space is rarely a restriction for them. Soil bacteria can number in the millions or even billions in a gram of productive soil (a volume approximately that of a red kidney bean). It's worth noting that bacteria are the most abundant of the soil microbes. The microbial biomass contributed by the various groups is maybe more important than the quantities of the distinct soil microorganisms. Among the microbial groups, soil fungi are the ones who contribute the highest biomass. In fact, they are commonly recognized as the major decomposer microorganisms in the soil because of their substantial contribution to biomass. Walking over a grassland in Africa or Australia, or through a cornfield in the American Midwest, you might be surprised to learn that there are literally "tons" of bacteria beneath your feet. Intriguingly, a fungus identified in Michigan could be one of the planet's largest living organisms. Armillaria bulbosa, a fungus identified in the state of Michigan in the United States, may turn out to be the world's largest creature, or at least among the largest. The fungus was discovered growing amid the roots of hardwood trees in a forest by scientists. The fungus' minuscule, branching filaments (called hyphae) cover 14.8 hectares (37 acres) of ground. A careful genomic investigation revealed that the filaments represent one creature. As they grow through the soil, fungi usually radiate outward in a circular manner. In truth, the fairy rings of mushrooms (so named because ancient peoples believed they represented the routes of dancing fairies in the night) that can be found in lawns and golf courses are the outside limit of a budding fungus. The section of the Michigan fungus that scientists have been able to identify could weigh up to 100 tons, about the same as a blue whale. The importance of these vast volumes of microbial biomass in the soil is not only due to their high metabolic capacity, but also due to the incredible diversity of biochemical reactions attributed to the soil microbial population. It's important to remember that soil microorganisms interact not only with other members of their own species, but also with other microbial species. It's not uncommon to discover, for example, that plant materials degrade significantly more quickly in the presence of a mixed soil population than when one or more groups of soil bacteria are removed from the system. There are different degrees of trophic (feeding) life in the soil. The soil microbial population, which destroys plant, animal, and microbial corpses and acts as a food source for some of the layers above it, is at the bottom of the trophic levels. Soil protozoa, for example, devour many bacteria and even some fungal spores. These are then devoured by larger soil animals (nematodes, mites, etc.) who are then eaten by even larger animals (e.g., worms and insects). In the absence of external inputs such as fertilizers, nutrients pass through this microbial food web, which is at the heart of controlling soil fertility and plant productivity. Soil microorganisms' ability to degrade organic molecules and hence regenerate a supply of carbon dioxide for plants is possibly their most important global function.

***Bacteria*:** Bacteria are the most common soil organisms, with at least a million of these small one-celled organisms in every gram of soil. One of the most important services bacteria give for plants is the provision of nutrients. Organic matter releases nitrogen, sulfur, phosphorus, and trace elements in some species. Others decompose soil minerals, releasing potassium, phosphorus, magnesium, calcium, and iron in the process. Plant growth hormones, which encourage or inhibit root growth, are produced, and released by a variety of species. Several bacteria species convert nitrogen from a gas in the air to forms that can be used by plants, and then back to a gas. A few bacteria species fix nitrogen in legume roots, while others fix nitrogen without regard for plant relationship. Bacteria assist in the transfer of nitrogen from ammonium to nitrate and vice versa, depending on soil conditions. Other benefits that bacteria bring to plants include enhancing nutrient solubility, improving soil structure, combating root infections, and purifying soil.

***Actinomycetes*:** Actinomycetes (ac-tin-o-my-cetes) are fungi-like threadlike bacteria. Although they are not as numerous as bacteria, they play an important role in the soil. They, like bacteria, aid in the decomposition of organic debris into humus, which releases nutrients. Antibiotics are also produced to combat root infections. Many of these antibiotics are also used to treat illnesses in humans. When a biologically active soil is tilled, actinomycetes are responsible for the delicious, earthy odor.

***Fungi*:** In soil, fungi come in a variety of types, sizes, and shapes. Threadlike colonies develop in some species, while one-celled yeasts appear in others. Plants benefit from fungi that break down organic matter or release nutrients from soil minerals. Fungi colonize larger chunks of organic matter quickly and start the breakdown process. Plant hormones are produced by some fungus, whereas antibiotics such as penicillin are produced by others. Fungi species have even been found to trap hazardous plant-parasitic nematodes. Mycorrhizae (my-cor-ry-zee) are fungi that reside on or inside plant roots and help root hairs reach deeper into the soil. Mycorrhizae boost water and nutrient intake, notably phosphorus. They are especially useful in soils that have been degraded or are less fertile. Root-feeding nematodes are less likely to penetrate mycorrhizae-colonized roots because the pest cannot pierce the dense fungal network. Mycorrhizae also create hormones and antibiotics, which help to promote root growth and prevent disease. The fungi gain an advantage by consuming minerals and carbohydrates released by the plant roots in which they dwell.

***Nematodes*:** Only a few kinds of nematodes are detrimental to plants, and they are abundant in most soils. The harmless species devour bacteria, fungi, algae, protozoa, and other nematodes, as well as decomposing plant litter. Nematodes, like other soil predators, hasten the pace of nutrient cycling.

***Arthropods*:** Arthropods are soil organisms that are visible to the naked eye. Sowbugs, millipedes, centipedes, slugs, snails, and springtails are among them. The principal decomposers are these. Their job is to devour and shred big plant and animal waste particles. Some people bury waste, exposing it to other soil creatures who breakdown it further. Smaller soil creatures are preyed upon by some members of this group. Springtails are tiny insects that mostly feed on fungus. Their feces are high in plant nutrients that are released after decomposition by fungi and bacteria.

***Earthworms*:** Burrows dug by earthworms improve water infiltration and soil aeration. Earthworm tunneling allows fields to absorb water at a pace four to ten times faster than fields without worm tunnels. This lowers runoff, recharges groundwater, and aids in the storage of more soil water during dry periods. Vertical earthworm holes pipe air further into the soil, promoting deeper microbial nutrient cycling. Edwards, Clive, and Bohlen (Edwards, Clive, and Bohlen, 1996). Where there is no tillage, earthworms thrive. Deep and frequent ploughing can lower worm populations by as much as 90%. Tillage kills earthworms by drying the soil, burying the plant debris they eat, and increasing the likelihood of freezing. Tillage can also kill and tear up worms, as well as demolish vertical worm burrows. In the spring and fall, young worms emerge. When farmers are likely to be tilling the soil, they are most busy. Reduce or eliminate tillage (particularly fall tillage), avoid using a moldboard plow, reduce residual particle size (using a straw chopper on the combine), add animal manure, and cultivate green manure crops to improve earthworm numbers. Leaving as much surface residue as possible is useful all year.

Humus

Humus is organic matter that has been decomposed to the point where it is refractory to further breakdown by soil microflora and fauna. Although humus makes up just about 5% of the soil by volume, it is an important source of nutrients and provides vital textural properties for soil health and plant growth. Arthropods, termites, and earthworms eat humus, which helps to enrich the soil. The result, humus, is suspended in colloidal form in the soil solution and produces a weak acid that chelates the iron and aluminum atoms of silicate minerals, allowing it to attack them. Humus has a high cation and anion exchange capacity, which is many times larger than clay colloids on a dry weight basis. It also functions as a buffer against variations in pH and soil moisture, like clay. Humic acids and fulvic acids are significant components of humus, which start off as raw organic matter. Bacteria begin to feed on the remains when plants, animals, and microbes die, culminating in the development of humus through the production of extracellular soil enzymes. Only molecules composed of aliphatic and aromatic hydrocarbons, assembled, and stabilized by oxygen and hydrogen bonds, survive in the form of complex molecular assemblages collectively known as humus when the residues decompose. In the soil, humus is never pure because it combines with metals and clays to generate complexes that add to its stability and soil structure. While the structure of humus contains few nutrients (except for constitutive metals such as calcium, iron, and aluminum), it can attract and linking cation and anion nutrients, which can then be released into the soil solution in response to selective root uptake and changes in soil pH, a process critical for the maintenance of fertility in tropical environments. Lignin is a type of lignin that is difficult to breakdown and accumulates in the soil. It also combines with proteins, making it more resistant to degradation by bacteria, including enzymatic destruction. Plant-derived fats and waxes are much more resistant to degradation and can last thousands of years in soil, which is why they're used as tracers of former vegetation in buried soil layers. Because organic molecules attach to and are maintained by clay, clay soils frequently have higher organic content and last longer than soils without clay. Apart from scleroproteins, proteins generally disintegrate quickly, but when coupled to clay particles, they become more resistant to degradation. Clay particles, like other proteins, absorb microbe-exuded enzymes, lowering enzyme activity but protecting extracellular enzymes from destruction. The addition of organic matter to clay soils can make that organic matter and any extra nutrients inaccessible to plants and microorganisms for years, yet a study found that adding mature compost to a clay soil boosted soil fertility. Nitrogen can be sequestered as resistant tannin-protein complexes in soils with high tannin concentration. The amount of plant material supplied each year and the type of base soil play a role in humus production. Climate and the kind of creatures present have an impact on both. The nitrogen level in humus-rich soils varies, although it is usually between 3 and 6 percent. As a source of nitrogen and phosphorus, raw organic matter is an important factor in soil fertility. Humus absorbs water and stretches and contracts more than clay between dry and wet states, enhancing soil porosity. As a result of microbial breakdown, humus is less stable than the soil's mineral ingredients, and its concentration decreases over time without the addition of fresh organic matter. Humus, on the other hand, in its most stable forms, can last for decades, if not millennia. Charcoal is a source of black carbon, a highly stable humus that has traditionally been used to improve the fertility of nutrient-poor tropical soils. This very old process, which can be traced back to the formation of Amazonian dark earths, has been revived and popularized under the name biochar. Biochar has been proposed to trap more carbon in order to combat the greenhouse effect.

SOIL FORMATION

Biological, climatic, geologic, and topographic forces all play a role in the evolution of soils. Soil formation is the evolution of soils and their qualities, and pedologists have identified five basic soil formation processes that influence soil properties. Parent material, geography, climate, organisms, and time are the five "state factors."

Parent Material

The earliest state of the solid stuff that makes up a soil is known as parent material. It can be made up of consolidated rocks as well as unconsolidated deposits including river alluvium, lake or marine sediments, glacial tills, loess (small, wind-deposited particles), volcanic ash, and organic matter (such as accumulations in swamps or bogs). The mineralogical content, texture, and stratification of parent materials all have an impact on soil formation (occurrence in layers). Dark-colored ferromagnesian (iron- and magnesium-containing) rocks, for example, can produce soils rich in iron compounds and clay minerals from the kaolin or smectite groups, whereas light-colored siliceous (silica-containing) rocks produce soils rich in iron compounds and clay minerals from the illite or vermiculite groups. Granitic rocks have a coarse texture that results in a coarse, loamy soil texture that supports the development of E horizons (the leached lower regions of the topmost soil layer). Basaltic rocks, on the other hand, have a fine texture that produces loam or clay-loam soils and prevents the development of E horizons. On coarse parent material, clearly defined E horizons tend to form more fully because water percolates to greater depths and drains more easily through coarse texture soils. In theory, parent material is either newly exposed solid matter (for example, volcanic ash soon after ejection) or deep-lying geologic material separated from atmospheric water and life. In practice, parent materials might be continuously deposited by wind, water, or volcanoes, and they can be altered from their solitary state, making identification difficult. The soil is monogenetic if a single parent material can be identified for the full profile; otherwise, it is polygenetic. Soils formed atop sedimentary rocks or unconsolidated water- or wind-deposited materials are examples of polygenetic soils. As in southeastern England, where soils produced atop chalk bedrock layers are overlain by soil layers developed on both loess and clay materials that have been changed by chalk dissolving, these so-called stratified parent materials can give soils with intermixed geologic layering and soil horizons. Because of the differences in parent materials, adjacent soils frequently have varied profile features. Lithosequences are the names given to the various soil areas, and there are two sorts of lithosequences. Soils developed on loess deposits at increasing distances downwind from their alluvial source are an example of continuous lithosequences with parent materials whose characteristics vary progressively over a transect. With increasing distance from the source, areas of similar deposits in the central United States or China show consistent declines in particle size and rate of deposition. As a result of the weathering of the loess particles, they reveal increases in clay content and the level of profile development. Discontinuous lithosequences, on the other hand, result from abrupt changes in source material. One soil created on schist (a silicate-containing metamorphic rock rich in mica) and another soil formed on serpentine is a basic example (a ferromagnesian metamorphic rock rich in olivine). In unconsolidated parent materials, more subtle discontinuous lithosequences, such as those found on glacial tills, display regular change in mineralogical composition or texture.

Topography

When considering topography as a soil-forming component, the geologic structural qualities of elevation above mean sea level, aspect (a landform's compass orientation), slope shape (i.e., either convex or concave), and relative position on a slope are all considered (that is, from the toe to the summit). Topography has an impact on how the hydrologic cycle affects earth material, particularly in terms of runoff and evapotranspiration. Precipitation can either wash off the ground surface, causing soil erosion, or it can percolate into soil profiles and become part of subsurface runoff, eventually entering the stream system. On a convex slope right below the summit, erosive runoff is most likely, but lateral subsurface runoff tends to generate a deposit of soluble or suspended debris towards the toe-slope. Lower elevation and an equatorially oriented aspect favor the conversion of precipitation to evapotranspiration. Topo-sequences are groups of soils that have different profile features due to the influence of local topography. Soil profiles on convex upper slopes in a topo-sequence are often shallower and have fewer distinct subsurface horizons than soil profiles at the summit or on lower, concave-upward slopes. Organic matter content, as well as clay content and soluble chemical concentrations, tend to rise from the summit to the toe-slope. Subsurface runoff is frequently the most prominent consequence of topography (or drainage). In humid temperate regions, well-drained soil profiles near a summit may have thick E horizons (leached layers) overlying well-developed clay-rich Bt horizons, whereas poorly drained soil profiles near a toe-slope may have thick A horizons overlying extensive Bg horizons (lower layers whose pale color signals stagnation under water-saturated conditions). These profile traits give way to less distinct horizons in humid tropical locations with dry seasons, with accumulations of silica, manganese, and iron near the toe-slope, but in semiarid regions, soils near the toe-slope accumulate soluble salts sodium chloride or calcium sulfate. The reality that topography is subject to significant changes over time tempers these broad findings. A-horizons are removed by water or wind erosion, exposing B horizons to weathering. The combined action of water and gravity can cause large chunks of entire soil profiles to migrate downslope abruptly. Natural disasters such as volcanic eruptions, earthquakes, and destructive storms can have apparent effects for geomorphologic patterns' instability.

Climate

In pedology, the term climate refers to meteorological features as they change across time periods longer than those required for soil attributes to develop. Precipitation, temperature, and storm patterns, both averages and variations, are among these features. Soil development is primarily influenced by climate through the actions of water and solar radiation. Water is the solvent in which chemical reactions occur in the soil, and it is necessary for soil organisms' life cycles. The primary medium for the erosive or percolative movement of solid particles is water. The amount of energy available from the sun determines the rates at which these water-mediated activities occur. Climate's integrated effects can be seen along a transect from pole to Equator on a global scale. Polar desert soils give way to extensively leached soils like the Podzols (Spodosols), which have an eye-catching ash-colored E horizon typical of humid boreal climates as one travels from the pole to cold tundra or forested regions. As climates warm further into temperate zones, organic matter accumulates in soils, and as evapotranspiration rises, lime (calcium carbonate) tends to concentrate closer to the top of the soil profile. The desert soils are deficient in organic matter and high in soluble salts, resulting in an arid subtropical climate. As the environment nears the Equator becomes more humid, high temperatures mix with high precipitation to produce red and yellow tropical soils, whose colors show the presence of residual iron oxide minerals that, due to their limited solubility, are resistant to leaching losses. A transect over the central United States from east to west highlights the effects of increased evapotranspiration on a continental scale. Soils with E horizons occur first, followed by soils with a lot of organic content. These give place to lime-rich soils, which eventually give way to desert soils with soluble salt efflorescence (powdery crust) at the top. Variations in climate can have a substantial impact on soil qualities on a regional scale, resulting in a contiguous sequence of soils known as a climo-sequence. A typical climo-sequence can be found along a 1,000-kilometer (600-mile) north-south transect in California's foothills of the Cascade and Sierra Nevada mountains. The profile properties of soils that have evolved on landscapes with similar topography fluctuate continually with differences in annual precipitation. The soils generated at the transect's dry southern end are shallow and stony, while those formed at the humid northern end have well-developed B horizons and are reddish in color. The smectite group, mixed vermiculite or illite group/kaolin group, and lastly the kaolin group alone respond to increased precipitation in the upper 20 cm (8 inches) of these soils. As soil leaching spreads deeper with increasing rainfall, these changes are mostly due to increased loss of silica and soluble metals. With increased precipitation, soil acidity and organic matter content rise, but readily soluble forms of calcium (essential for plant growth and soil aggregation) fall. In theory, climatic indicators might be regarded as soil profile traits that are closely linked to climate. A soil profile with two well-defined zones of lime accumulation, one shallow and one deep, for example, could indicate the presence of a past climate with higher precipitation levels that drove the lime layer deeper than the current climate can. Paleosols are soils that developed in different habitats from the current one and are preserved (at least partially) at greater depth. Some characteristics of these soils can be used as climate indicators, with horizons with hardened accumulations of relatively insoluble iron, manganese, or calcium minerals, or layers with accumulations of tightly aggregated clay-size particles being the most dependable. Pedologists may be able to deduce previous climate by knowing the clay mineral in a probable paleosol and assuming the precipitation-clay mineralogy relationship stated above. The depth of lime-containing layers in a paleosol can be used to infer the precipitation level of a former climate. To separate the impacts of previously weathered parent material from those of merely climatic influence, these possible applications of climatic relationships must be carefully studied.

Organisms

Vegetation, animal numbers, and human populations can all have an impact on soil development. A bio-sequence is a collection of contiguous soils impacted by local flora and fauna. To back to the climo-sequence outlined earlier, the vegetation seen throughout this narrow foothill region spans from shrubs in the dry south to needle-leaved trees in the wet north, with large grasslands in between. Transition zones exist in the middle of the precipitation range, where small groves of needle-leaved trees are randomly mixed with grassland areas. These plant populations exhibit a climate-selected native flora. The qualities of the soils beneath these plants, on the other hand, show variances that are caused by the different plant species rather than climate, geography, or parent material. The soils under trees, for example, are significantly more acidic and contain much less humus than those under grass, and the grassland soil has a much higher nitrogen content. These characteristics are derived directly from the types of litter created by the two types of vegetation. Relatively young soils generated from alluvial parent material frequently provide a chance to study bio-sequences. This type of soil, which is found beneath shrubs, may be richer in humus and plant nutrients than soil found beneath needle-leaved trees. Differences in the cyclic processes of plant growth, litter formation, and litter decomposition cause this variation. If litter production is low, organic matter decomposers will feed on stored material in the soil, whereas high litter production will allow soil organic matter stocks to increase, resulting in humus-rich A horizons rather than the leached E horizons found in soils formed under humid climatic conditions. Humans play a role in the biological influx that affects soil development. Human influence can be as drastic as the wholesale removal or burial (by urbanization) of an entire soil profile, or as subtle as agriculture's progressive change of organic matter or irrigation's gradual modification of soil structure. Cultural practices have a big impact on the chemical and physical qualities of soils that are important for crop growth. Loss of arable land, erosion, salt building, and organic matter depletion are only a few of the issues that cultural traditions have generated for agriculture.

Time

The soil-forming elements of parent material and topography are primarily site-related (terrain characteristics), whereas climate and organisms are primarily flux-related (inputs from the surroundings). Time is neither a terrain attribute nor a source of external stimuli when it comes to soil formation. Instead, it's an abstract variable whose primary purpose is to track the evolution of soil characteristics. Time's conceptual independence from its four partner variables simply means that soil evolution can take place while site characteristics and external inputs remain mostly constant. Certain soil profile characteristics can be viewed as markers of time. (A chrono-sequence is a set of soil profiles that differ only in terms of age.) One example of a time-related feature is the humus content of the A horizon, which, for soils younger than 10,000 years, rises at a rate that is dependent on parent material, vegetation, and climate. After around 10,000 years, this rate of increase slows, plant nutrients begin to seep away, and soils approaching one million years show a considerable reduction in humus content. (Agricultural techniques can halt this development, resulting in a 25 percent or more decrease in stored humus, as well as changes in economically significant soil qualities including nutrient availability and retention, acidity, and aeration.) Aging is also indicated by the accumulation of clay and lime in soil profiles as a result of their downward transfer. Older soils created on calcium-rich loess deposits, for example, have more developed E and Bt horizons (as well as narrower A horizons) than younger soils formed on same deposits. Similarly, soils grown on alluvium can acquire a clayey hardpan after 100,000 years or so in a chrono-sequence. Regardless of climatic conditions, soils tend to redden in hue as they age, indicating the survival of weakly crystalline or crystalline oxide minerals containing Fe3+. Indeed, the prevailing mineralogy of clay-size particles in soils can be used to determine the age of the soil. The set of Jackson-Sherman weathering stages refers to any sequence of predominant clay mineralogy found in a soil (see the table). Each step down the table corresponds to a longer period of mineral residence time, both between and within the three main stages (early, intermediate, and advanced).

Table 5: soil weathering stages

	characteristic minerals in soil clay fraction	characteristic chemical and physical conditions of soil	characteristic soil profile features
early stage	gypsum, carbonates, olivine/pyroxene/ amphibole, Fe^{2+}-bearing micas, and feldspars	low water and humus content, limited leaching, reducing environments, and limited time for weathering	minimally weathered soils all over the world, though mainly in arid regions where low rainfall keeps weathering to a minimum
intermediate stage	quartz, mica/illite, vermiculite/chlorite, and smectite	retention of Na, K, Ca, Mg, Fe^{2+}, and silica; alkalinity and ineffective leaching; igneous rock rich in Ca, Mg, and Fe^{2+} but no Fe^{2+} oxides; easily hydrolyzed silicates; and transport of silica into the weathering zone	soils of temperate regions developed under grass or trees—i.e., the major agricultural soils of the world
advanced stage	kaolin, aluminum and iron oxides, and titanium oxides	removal of Na, K, Ca, Mg, Fe^{2+}, and silica; effective leaching by fresh water; low pH; and dispersion of silica	intensely weathered soils of the humid tropics, frequently characterized by acidity and low fertility

Source: Jackson & Sherman, (1953).

The prevalence of sulfates, carbonates, and primary silicates in the soil clay fraction, other than quartz and muscovite, indicates early weathering. These minerals can only live if soils are mostly dry, cold, or wet; in other words, if they have little exposure to water, air, or solar energy. The clay fraction in the intermediate stage is dominated by quartz, muscovite, and secondary aluminosilicates. These minerals may withstand leaching conditions that do not deplete silica and metals like calcium and magnesium, as well as those that do not result in total oxidation of Fe2+, which is subsequently integrated into illite and smectite clays. The advanced stage, on the other hand, is marked by extensive leaching and harsh oxidizing conditions, leaving only aluminum, Fe3+, and titanium oxides. Only if the removal of silica by leaching is incomplete, or if there is an encroachment of silica-rich waters, as can happen when water percolating through a soil profile at the upper part of a topo-sequence moves downslope into a soil profile at the lower part, will kaolin become the dominant clay mineral group. The inference that it takes longer time to develop soils with the more permanent minerals in the clay fraction is implicit. This result is supported by extensive field research conducted around the world to assess the rate of soil horizon formation. It takes less than a decade for soils with early clay mineralogy to produce a centimeter (0.4 inch) of horizon thickness. Soils with dominant clay minerals in the intermediate stage can create a centimeter of solum in less than a century, whereas soils with dominant clay minerals in the advanced stage can take hundreds of years.

SOIL CLASSIFICATION

The soil order system of the United States Soil Taxonomy and the soil group system, published as the World Reference Base for Soil Resources, devised by the United Nations Food and Agriculture Organization (FAO), are the two main systems of soil classification in use today. Both systems are morphogenetic in that they classify soils based on structural characteristics while also considering the five elements of soil formation discussed in the previous section when deciding which characteristics to prioritize. Diagnostic horizons, well-defined soil layers whose structure and origin may be attributed to soil-forming processes and used to distinguish among soil units at the highest level of classification, are central to both systems (see the table of primary diagnostic horizons). Diagnostic horizons can be located extremely close to the ground surface (epipedons) or very deep in the soil profile (subsurface horizons); they don't have to match the horizon letter designations.

Table 6: Primary diagnostic horizons of soil

	U.S. Soil Taxonomy	defining features	FAO soil group system
epipedons	histic	thick organic layer	histic
	mollic	thick, dark, neutral to alkaline	mollic
	ochric	pale or thin	ochric
	umbric	thick, dark, acidic	umbric
subsurface horizons	argillic	clay mineral deposition	argic
	cambic	in situ mineral weathering only	cambic
	oxic	highly weathered; aluminium oxide, iron oxide, and kaolin clay deposition	ferralic
	spodic	aluminium oxide, iron oxide, and humus deposition	spodic

Source: https://www.britannica.com/science/soil/Soil-classification

The presence of a characteristic horizon in a soil profile is frequently enough to identify its taxonomic class at the order (US) or group level (FAO). Soil profiles having mollic epipedons, for example, are classified as Mollisols in the US Soil Taxonomy. Mollic A horizons, on the other hand, are found only in FAO soil categories whose features are influenced by a steppe environment (that is, Chernozem, Kastanozem, and Phaeozem). Both the US and FAO titles refer to soils that have evolved in plains under grassland vegetation and have a high humus content in the A horizon due to widespread root growth. When soil maps of the United States based on the US and FAO taxonomies are examined, however, the relationship between the two taxonomic systems is not always as tight as in this example.

U.S. Soil Taxonomy

The United States Soil Taxonomy divides soils into six types. Here, only the top-level category, order, is examined. As shown in the table of the US Soil Taxonomy, soil orders are named by adding the suffix -sol to a root word. The 12 soil order names that arise represent a classification based on parent material or processes associated with the five components of soil formation as represented in diagnostic horizons.

Table 7: U.S. soil taxonomy

soil order	defining characteristics	name derivation	percent of Earth's land area*
Alfisol	moderate leaching; B horizon enriched in clay; humid forest vegetation	Pedalfer (C.F. Marbut)	9.77
Andisol	volcanic-ash parent material	a do (Japanese: "dark soil")	0.73
Aridisol	hot, dry climate; weak B horizon	aridus (Latin: "dry")	18.53
Entisol	little or no horizonation or swelling clay	recent	10.61
Gelisol	permafrost within 2 metres (approximately 6 feet) of the land surface	gelid (Greek: "very cold")	—
Histosol	organic parent material	histos (Greek: "tissue")	1.84
Inceptisol	little or no B horizon development	inceptum (Latin: "beginning")	21.80
Mollisol	thick, soft, black A horizon	mollis (Latin: "soft")	5.99
Oxisol	hot, humid climate; B horizon enriched in iron and aluminum oxides and kaolinite	oxide (French)	7.00
Spodosol	cool, humid climate; B horizon enriched in iron and aluminum oxides and organic matter; sandy parent material	spodos (Greek: "wood ashes")	3.45
Ultisol	warm, humid climate; B horizon enriched in clay; extensive leaching	ultimus (Latin: "last")	8.12
Vertisol	little or no horizonation; high content of swelling clay	vertere (Latin: "to turn")	2.23

Source: Hallberg, (1984).

The soil orders associated with certain types of parent material (Andisol, Histosol, and Vertisol) cover fewer than 5% of the Earth's continental land area. Nearly a third of the land surface is covered by soils that have not developed because they are too young (Entisol) or because they are in an undesirable weathering environment (Inceptisol). Nearly 40% of the total is made up of soils that are likely to induce natural toxicity to agricultural plants due to salt accumulations (Aridisol) or acidity and aluminum (Spodosol, Oxisol, and Ultisol). As a result, the Alfisols and Mollisols, which account for roughly 15% of the total land area, are the world's fundamentally more productive soils. They occupy a strategic zone in the Northern Hemisphere and South America at middle latitudes.

FAO Soil Groups

The FAO's categorization method is based on a two-level nomenclature that includes the name of a soil group and a modifying adjective that is used to designate a soil unit within that group on the FAO Soil Map of the World. It is not intended to replace national soil categorization systems such as the United States Soil Taxonomy, but rather to make comparisons between them easier. Only the major soil groups are covered in this article. Four of the soil groups are primarily defined by their parent material (first cluster in the FAO classification system table), four are mostly related to topographic variables in soil formation, and the remaining 22 are based on the three additional soil-forming elements: climate, organisms, and time. The FAO system's soil groups, like the US soil orders, are based on vast sets of field and laboratory observations as well as technical criteria.

Table 8: Soil classification system of the Food and Agriculture Organization

	soil group	**abbreviation**	**defining characteristics**	**name derivation**	**percent of Earth's land area**
Soils defined by parent material	Andosol	AN	volcanic ejects	An do (Japanese: "dark soil")	0.88
	Arenosol	AR	sands	arena (Latin: "sand")	7.17
	Histosol	HS	organic matter	histos (Greek: "tissue")	2.51
	Vertisol	VR	swelling clays	vertere (Latin: "to turn")	2.67

Soils defined by topography	Fluvisol	FL	alluvial low-lands	fluvius (Latin: "river")	2.79
	Gleysol	GL	waterlogged lowlands	gley (Russian: "mucky soil mass")	5.74
	Leptosol	LP	eroded uplands	leptos (Greek: "thin")	13.19
	Regosol	RG	climate-limited, thin soil	rhegos (Greek: "blanket")	2.07
Soils defined by climate, organisms, and time	Calcisol	CL	calcium car-bonate accumu-lation	calix (Latin: "lime")	6.38
	Gypsisol	GY	gypsum accu-mulation	gypsum (Lat-in: "calcium sulfate")	0.72
	Solon-chak	SC	salt accumula-tion	sol chak (Rus-sian: "salty area")	2.55
	Solonetz	SN	sodium accu-mulation	sol etz (Rus-sian: "strong-ly salty")	1.08
	Durisol	DU	silica accumu-lation	durum (Latin: "hard")	—
	Cherno-zem	CH	cold steppe environment	chern zem-lja (Russian: "black earth")	1.83
	Umbrisol	UM	cool, wet steppe environ-ment	umbra (Latin: "shade")	0.80
	Kastano-zem	KS	warm, dry steppe environ-ment	castanea zemlja (Latin-Russian: "chestnut earth")	3.71

	Phaeozem	PH	warm, wet steppe environment	phaios zemlja (Greek-Russian: "dusky earth")	1.51
	Acrisol	AC	seasonally dry humid tropics	acer (Latin: "strong acid")	7.97
	Alisol	AL	humid subtropical and warm temperate areas	alumen (Latin: "aluminum")	0.80
	Ferralsol	FR	extensively weathered; humid tropics	ferrum alumen (Latin: "iron-aluminum")	5.98
	Lixisol	LX	driest humid tropics	lixivia (Latin: "washing")	3.47
	Nitisol	NT	extensive clay migration; tropics	nitidus (Latin: "shiny")	1.59
	Plinthosol	PT	fluctuating water table; plinthite	plinthos (Greek: "brick")	0.48
	Luvisol	LV	clay accumulation; distinct seasons	luere (Latin: "to wash")	5.18
	Planosol	PL	clayey horizon	planus (Latin: "flat")	1.04
	Podzol	PZ	accumulation of iron and aluminum oxides and humus	pod zola (Russian: "under ash")	3.87
	Albeluvisol	AB	cold temperate area; bleached horizon over clayey horizon	albus (Latin: "white")	2.55

	Cryosol	CR	alternate freezing and thawing; water-logged during thaw; permafrost within 1 metre (3 feet) of the land surface	kryros (Greek: "cold")	—
	Anthrosol	AT	extensive human modification	anthropos (Greek: "man")	0.004
	Cambisol	CM	little soil formation; recent	cambiare (Latin: "to change")	11.96

Source: FAO 1996

Some of the FAO soil groups are very similar to soil orders in the United States (for example, Andosol, Cambisol, Histosol, and Vertisol). Others are more closely related to lower levels of nomenclature than the soil order; for example, Gypsisol, Calcisol, Solonchak, and Solonetz are typically categorized inside the Aridisol order in the United States. Others have no counterpart in the United States' taxonomy (for example, Anthrosol). The FAO classifies eight soil categories as having high inherent soil fertility: Cambisol, Chernozem, Fluvisol, Gleysol, Kastanozem, Phaeozem, Umbrisol, and Vertisol. They account for 31% of the overall land area. If the Cambisol and Vertisol groups are eliminated, the percentage drops to 16%, which is quite like the estimate stated above for the more fertile U.S. soil orders. According to the FAO, the soil types that pose toxicity risks due to salt buildup (Calcisol, Gypsisol, Solonchak, and Solonetz) or aridity and aluminum accumulation (Acrisol, Alisol, and Ferralsol) cover roughly 25% of the land area. If other tropical soil types, as well as the Podzol and Albeluvisol groups, were included, the percentage would rise to over 33%. As a result, all soil categorization systems come to the same conclusion: the naturally more fertile soils make up a small part of the total soil resources on Earth.

Chapter 2

Soil Degradation

CONTENTS

INTRODUCTION

In general, over the last four decades, increased provisioning of food, fuel, and fiber has resulted in soil deterioration and the loss of numerous supporting and regulatory services given by soils. The loss of soil characteristics and regulating ecosystem services will have an influence on ecosystem provisioning services in the long run. One of the frontiers of soil science is determining the factors that influence the stability and resilience of soils in the face of disturbance.

Soil degradation is described as a deterioration in soil health that causes the ecosystem's ability to offer goods and services to its beneficiaries to be harmed. Degraded soils have a state of health that prevents them from providing the regular products and services that the soil provides in its ecosystem. These definitions, which are mostly based on the LADA concept of land degradation, are critical in capturing the complexities of degradation processes and their subjective assessment by various stakeholders in soil and land. The following are some popular soil degradation definitions:

- **Soil Erosion:** Soil erosion is a word that is frequently conflated with soil degradation, but it solely refers to absolute topsoil and nutrient losses. Although this is the most obvious result of soil deterioration, it does not encompass all of it. In mountainous places, soil erosion is a natural occurrence, although it is frequently exacerbated by inadequate management techniques.
- **Land degradation:** Land degradation encompasses any negative changes in the ecosystem's potential to provide goods and services (including biological and water-related goods and services – and, in LADA's perspective, also land-related social and economic goods and services).
- **Desertification:** Another word for (a) land deterioration in dryland areas and/or (b) the irreversible alteration of land to a state where it can no longer be restored for its original purpose is desertification.
- **Prevention:** The employment of conservation methods to protect natural resources and their environmental and productive value is referred to as prevention.
- **Mitigation:** Mitigation is a strategy for halting the progression of environmental damage. This arrives at a time when the deterioration process has already begun. The major goal is to

stop future degradation and begin to improve resources and their functions. The effects of mitigation activities are usually visible in the short to medium term, which gives a strong incentive to continue working. The term "mitigation" is also used to denote the lessening of degradation's effects.

- **Rehabilitation:** When the land has degenerated to the point where the original use is no longer practicable and the land has become virtually unproductive, rehabilitation is necessary. To show any benefit, longer-term and generally more expensive expenditures are required.

TYPES AND PROCESS OF SOIL DEGRADATION

While soil deterioration occurs naturally, manmade activities have greatly accelerated it. Furthermore, soil degradation is worsening as a result of climate change and human activity. Overgrazing, deforestation, poor land management, fuelwood harvesting, and urbanization are the five major anthropogenic drivers of soil degradation worldwide, in order of increasing magnitude. Deforestation, biomass burning, nutrient depletion, and overgrazing almost always result in the loss of natural vegetative cover, which causes soil deterioration. Rainfall, which breaks soil aggregates, and higher temperatures, which enhance soil organic matter breakdown rates, expose the soil surface; in addition, litterfall and roots, the principal sources of organic inputs that maintain soil organic matter, are removed, or greatly reduced. The kind and intensity of land use determines subsequent rates and types of soil degradation. Soil degradation can happen quickly depending on how management strategies, beginning soil conditions, vegetation, and environmental factors like rainfall interact. Physical, chemical, and biological processes are used to classify soil degradation; this classification helps to identify relationships between land management, degradation processes, and soil processes.

Soil Physical Degradation

Physical degradation takes many different forms, including deformation of the inner soil structure caused by compaction, which is caused by heavy agricultural machinery tracking, vertical sheer stress caused by rotating parts of machines, such as wheels, and chemical impacts such as salinization, particularly alkalinization. Soil erosion by water and wind, as well as the production of crusts at the soil surface, are examples of physical degradation (soil crusting). Heavy soil loading causes deformation of the inner soil structure, which includes large (>50 mm diameter) and small macropores (50–10 mm diameter), medium pores (10–0.2 mm diameter), and fine pores (0.2 mm diameter), and results in the loss of soil macropores >50 mm diameter or even smaller macropores >10 mm diameter, especially under extreme conditions. Macropores are responsible for soil aeration and water conductivity, as well as providing living space for soil creatures, particularly those that cannot form their own spaces in the soil, such as many earthworms. As a result, the loss of macropores, especially medium pores (10–0.2 mm in diameter), results in reduced water (rainwater) conductivity in soils, resulting in stagnant water phases and anaerobic conditions that alter soil chemistry and life. Compaction also causes soil erosion by water surface flow under certain topographical circumstances. Sheer stress has similar effects, causing smearing and closing or clogging pores, especially in the top soil, between 0 and 20 cm in depth. The formation of so-called plough horizons, a compacted zone below the ploughed top soil with problems for root penetration and often changes in redox potential due to decreased water infiltration, causing stagnant water conditions after heavy rainfall, is a unique form of sheer stress combined with loading and compaction. Compaction at soil levels greater than 50 cm is nearly permanent because no technical solutions exist to restore soil macro-porosity, apart from biological processes, such as earthworm activity, which is normally a lengthy process that is dependent on subsequent building circumstances. Furthermore, physical soil degradation occurs at the soil surface through the destruction of crumby or blocky soil structures, primarily due to the physical impact of rainwater, particularly raindrops with high kinetic energy, known as splash, which destructs surface soil structures, causing smaller soil particles to be transported by water or wind (erosion) or form crusts at the soil surface. The amount of physical damage caused by high-kinetic-energy raindrops varies on the size of the droplets, the wind speed, and the type and stability of surface soil structures. Soil organic matter has a big impact on structure stability. Soil structures become instable and hence prone to erosion as soil organic matter un top soils depletes, especially crumbs or small blocks at the soil surface. Wind and water erosion pose the greatest threat to soils with a high silt concentration. Soil erosion results in the loss of the most fertile part of the soil, as well as a reduction in soil fertility due to a reduction in rooting depth, and hence the availability of water and nutrients for plant growth. Fine soil particles formed by rain splash at the soil surface can create crusts in climatic zones with lengthy dry intervals, which are usually particularly resistant to deformation by water and wind, as well as fungal development compacting these surfaces. Crusting is especially common in soils with a Mediterranean climate or in semiarid or semi-humid tropics and subtropical climates. Nearly all soil features and processes, including as space for plant roots and soil biota, soil temperature, water, air, and nutrient transport, and natural attenuation of organic and inorganic contaminants, are negatively impacted by physical soil deterioration.

Soil Chemical Degradation

Excessive buildup of nutrients (eutrophication), salts (salinization in the root zone and beyond), or toxic materials are associated with soil chemical imbalances caused by a chemical reaction or pH; declines in the availability of plant nutrients (nutrient depletion); and excessive buildup of nutrients (eutrophication), salts (salinization in the root zone and beyond); and excessive buildup of nutrients (eutrophication). Nutrient depletion, also known as soil fertility decrease, is the most common form of chemical degradation in much of the tropics, particularly Africa, when nutrient losses owing to crop waste removal, harvested products, erosion, and leaching are not adequately supplied by external inputs. Reduced crop yield and biomass in general, as a result of nutrient depletion, leads to additional decreases in soil organic matter. Ultisols, Oxisols, and sandy Inceptisols are soils with low initial nutrient capital, low cation exchange capacity, low activity variable charge clays, and low soil organic matter that deplete more quickly than soils without these traits. As has been done for soil erosion and productivity losses, there is a growing corpus of material that will be valuable in establishing linkages between nutrient depletion and reduced plant output (Sanchez & Jama, 2002). Soil eutrophication, on the other hand, is a degrading process that occurs predominantly in industrialized countries in temperate regions where large-scale agriculture uses excessive amounts of fertilizer, manures, and pesticides.

Soil Biological Degradation

Soil organic matter and soil biota underpin many important soil processes, hence biological degradation is frequently associated with a decrease in soil organic matter and a loss of soil biota. The most extensively researched form of biological degradation is the decrease of soil organic matter when natural systems are converted to agriculture and when agriculture is intensified through tillage.

Table 9: Types of soil degradation and causes and impacts on soil processes

Type	Causes (not one to one along row)	Degradation process	Impact on soil processes
Physical	Deforestation	Breakdown of soil structure, aggregation, and porosity	Reduction in infiltration capacity Changes in soil water-retention characteristics
	Biomass burning	Crusting and surface sealing	Increase in runoff rate and amount
	Tillage up and down slope, excessive animal, human, and machine traffic, overgrazing	Compaction of surface and subsoil, reduction in proportion and strength/stability of aggregates	Accelerated erosion by water and wind Increase in bulk density leading to reduction in porosity Water logging and anaerobiosis
Chemical	Irrigation with poor quality water, inadequate drainage	Salinization, alkalinization	Accumulation of base-forming cations
	Little to no use of fertilizers	Nutrient depletion	Decreased levels of macronutrients on exchange sites, soil organic matter, and in soil solution
	Excess use of fertilizers	Acidification, eutrophication	Leaching and runoff of nutrients to water sources
	Application of industrial, urban wastes	Toxification, contamination with heavy metals, pollution	Excessive buildup of some elements (e.g., Al, Mn, Fe) and heavy metals (e.g., lead and mercury); increase in soilborne pathogens
Biological	Removal of or burning residues	Depletion of soil organic carbon	Reduction in N mineralization, soil aggregation, and related properties
	Little or no use of organic inputs	Decline in diversity and abundance of soil biota	Shift in species composition and diversity of favorable soil organisms
	Monoculture, excessive tillage	Loss of soil structure	Reduction in porosity and infiltration, reduction in activity of soil biota

Source: Lal, et al., (2003).

The rate of change in soil organic matter content, as well as the magnitude of change, are influenced by soil type (which is slower in clayey soils), land use, and climate (slower in colder or drier climates and waterlogged conditions). The corpus of literature on soil carbon changes when natural systems are transformed to annual croplands is large and sufficient to give the pedo-transfer functions needed to link soil property loss to a variety of ecosystem processes. Other land-use transitions, such as natural systems to pastures or tree plantations, annual cropping systems to pastures or tree-based systems, or even changes in annual cropping system management, have produced more recent data. In both temperate and tropical soils, a meta-analysis of soil carbon changes with land-use change shows a 50 percent decline in the top 30 cm when forests were converted to cropland, a 15 percent decline when forests were converted to coniferous plantations, no decline when forests were converted to broadleaf plantations, and an overall increase of about 10% when forests were converted to broadleaf plantations.

CAUSES OF SOIL DEGRADATION

Natural risks, direct causes, and underlying causes are the three types of causes for land degradation. Natural hazards are physical environment conditions that result in the presence of a high degradation hazard, such as steep slopes as a hazard for water erosion. Unsuitable land use and land management techniques, such as the cultivation of steep slopes without soil conservation measures, are direct causes. The reasons for these improper types of land use and management include underlying causes; for example, the slopes may be cultivated because the landless poor require food, and conservation measures may not be implemented because these farmers lack tenure security. There is a contrast between unsuitable land use and incorrect land management approaches, while there is some overlap. Unsuitable land use refers to the use of land for purposes that are not environmentally viable. Forest removal and arable usage of steeply sloped upper watershed lands that would be more valuable to the community as water supplies if managed under a protected forest cover are two examples. Inappropriate land management methods refer to the use of land in ways that, if properly managed, could be sustainable, but where the necessary procedures are not followed. Failure to implement soil conservation measures when they are required is one example. It can also apply to land use that is ecologically sustainable at low intensities but becomes ineffective at greater intensities due to poor management. Shifting cultivation and semi-arid rangeland grazing are two examples. Only four causes were identified in this analysis, and they are as follows:

- deforestation and removal of natural vegetation;
- overgrazing;
- agricultural activities;
- Over-exploitation of vegetation for domestic use.

Natural Soil Degradation Hazards

The following are the principal natural dangers, which function as predisposing factors for land degradation:

For water erosion:

- rains of high intensity;
- steep slopes of the mountain and hill lands;
- soils with low resistance to water erosion (e.g., silty soils, vertisols).

For wind erosion:

- semi-arid to aria climates;
- high rainfall variability, with liability to drought spells;
- soils with low resistance to wind erosion (e.g., sandy soils).
- an open cover of natural vegetation.

For soil fertility decline:

- strong leaching in humid climates;
- soils which are strongly acid, and/or with low natural fertility.

For waterlogging:

- alluvial plains or interior basins which restrict outward drainage of groundwater.

For salinization:

- semi-arid to aria climates with low leaching intensity;
- plains and interior basins which restrict outward drainage of groundwater;
- soils which are naturally slightly saline.

For lowering of the water table:

- semi-arid to aria climates with low rates of groundwater recharge.

In some circumstances, these natural risks are severe enough to result in unproductive land without the intervention of humans. Natural saline soils found in some interior basins of dry climates, or areas of natural gullying ('badlands') are examples. Problem soils are the name given to such situations. In Dent, the percentage of land covered by issue soils is indicated (1986).

Direct Causes of Degradation

***Deforestation*:** Unsuitable terrain is deforested. Deforestation is a sort of degradation in and of itself, as well as a source of other types of degradation, most notably water erosion. Deforestation is not always detrimental; without it, the most productive agricultural fields (in both the temperate and tropical zones) would be unavailable. When the cleared area is steeply sloped or contains shallow or easily erodible soils, deforestation becomes a cause of degradation; and when the clearing is not followed by effective management, deforestation becomes a cause of degradation.

***Overcutting of vegetation*:** Vegetation overcutting to acquire lumber, fuelwood, and other forest products, rural people cut down natural forests, woodlands, and shrublands. When the rate of natural regrowth is faster than the rate of cutting, it becomes unsustainable. This has occurred frequently in semi-arid areas, where fuelwood shortages are common. The loss of natural woody cover such as trees and shrubs is a major contributor to both water and wind erosion. It is mentioned as a cause for 98 percent of the area damaged by wind erosion in the GLASOD evaluation. It is also mentioned as a contributory cause of salinization in this assessment.

Shifting cultivation without adequate fallow periods: Shifting cultivation used to be a viable land use option when low population density allowed for long enough forest fallow periods to repair soil characteristics. It has become unsustainable because to population growth and the forced shortening of fallow times. Shifting cultivation is prevalent in northeast India's hill regions, where it contributes to water loss and soil fertility decrease.

***Overgrazing*:** Overgrazing occurs when livestock stocking intensifies beyond the carrying capacity of natural pastures. It results in a reduction in the quantity and quality of vegetation cover. This is a major contributor to not only wind but also water erosion in arid terrain. Degradation of the plant cover and erosion both result in a decrease in the organic master and physical characteristics of the soil, and hence in erosion resistance. Intensive grazing at the conclusion of the annual dry season or during droughts does not always result in degradation; the plants may rebound after subsequent rains. Degradation happens when vegetation and soil qualities do not restore to their prior state during periods of regular rainfall.

***Non-adoption of soil-conservation management practices*:** Water erosion control is required on all sloping slopes when arable land is used. Wind erosion control techniques are also required on lever land in dry environments. The following are some examples of soil conservation management practices:

- Biological techniques include maintaining a "round surface cover" of living plants or plant litter, as well as vegetative barriers such as contour hedgerows and grass strips, as well as windbreaks and shelterbelts.
- Terraces and other types of bank-and-ditch structures are examples of earth structures (bunds, storm drains, etc.).
- Maintaining soil resistance to erosion entails, first and foremost, maintaining the organic master of the soil and, as a result, aggregation, and structure.

There has been a lot of effort put into encouraging the use of such management approaches. In certain instances, these initiatives have yielded significant results. In others, staff and resources have been severely lacking, and recommended procedures have been poorly adopted. Extension services have failed to fully embrace the new shift in emphasis on soil conservation, which includes more use of biological methods, such as agroforestry, and a larger emphasis on farmer engagement and economic incentives. The standard of management, not the environment or the type of land use, is frequently the cause of degradation. The manufacturing of tee in Sri Lanka's hill country is a good example. Poorly maintained farms have bare soil between plants, which causes severe erosion even on steep slopes.

Extension of cultivation onto lands of longer potential and/or high natural hazards: These are also known as "fragile" or "peripheral" areas. The more productive, or high-potential, agricultural fields were historically the first to be settled. The extensive usage of areas with a longer potential, those that are less fruitful or have greater degradation hazards, has resulted from population growth. The following are examples of marginal lands:

- steeply sloping land;
- areas of shallow or sandy soils, or with laterite crusts;

- Cultivation of semi-arid lands, and grazing of the crier semi-arid areas, marginal to deserts.

Such land contributes significantly to the country's agricultural output. It is neither desirable nor practical to pull them out of production except in locations with the greatest environmental risks, such as upper watersheds. It must be acknowledged that if the resources of such lands are to be preserved, greater management standards are required. They are, unfortunately, frequently used by impoverished farmers.

Improper crop rotations: Farmers in some areas have embraced cereal-based, intensive crop rotations, centered mostly on rice and wheat, in place of the more balanced cereal-legume rotations that were previously seen, due to population expansion, land scarcity, and economic constraints. This is a contributing factor in the loss of soil fertility.

Unbalanced fertilizer use: Farmers try to preserve agricultural yields when soil fertility has diminished due to protracted cultivation or erosion. The application of fertilizer is the most common method for doing so. In the near term, nitrogenous fertilizer provides the most immediate and cost-effective yield response. The ratios of nitrogen to phosphorus, as well as nitrogen to all other nutrients, have been steadily increasing in the region. Deficits of other nutrients, such as sulphur and zinc, have been recorded in areas where phosphate deficits have been identified and treated using phosphatic fertilizer. In the medium term, the short-term solution of using just macronutrients, notably nitrogenous fertilizer, to counteract fertility reduction is causing a worse problem of nutritional imbalance. Longer yield responses to fertilizers are expected to be one of the outcomes.

Problems arising from planning and management of canal irrigation: On the large-scale canal irrigation schemes of the Indo-Gangetic plains, the development of salinization and waterlogging has been extensively documented. From the 1930s onwards, the use of water in excess of natural rainfall resulted in a gradual rise in the water table. Waterlogging occurs when the water table is close to the surface, resulting in salinization due to evaporation of salt-containing water. In the soil exchange complex, sodium substitutes other bases, resulting in sodification. If deep drains had been incorporated in the early implementation of irrigation schemes, the problem could have been avoided or lessened. In development planning studies of Pakistani and Indian irrigated areas, more thorough details of the complicated processes involved can be found.

Over-pumping of groundwater: Tube-well technology has led to water abstraction in locations with non-saline ('sweet') groundwater in excess of natural recharge by rainfall and river seepage. As a result, the water table has gradually dropped, causing soil degradation.

Underlying Causes of Degradation

The factors for land degradation described above are merely the tip of the iceberg. They apply to all direct causes, apart from large-scale irrigation scheme difficulties that resulted from a lack of planning and management foresight.

Land shortage: Land has always been recognized as a finite resource, but the full significance of this reality has only recently been apparent. Previously, food scarcity or poverty might be alleviated by cultivating new, uncultivated land. This option is no longer available in most of South Asia. For India, Pakistan, Sri Lanka, and Afghanistan, the percentage change in agricultural land between 1980 and 1990 was less than 2.5 percent, while Bangladesh saw a tiny absolute decline. The growth in Nepal has undoubtedly been achieved by deforestation and the agricultural usage of sloping land that is difficult to farm on a long-term basis. Land scarcity has resulted in reductions in already tiny areas of agricultural land per person in six of the eight countries, including all in the humid zone, when paired with increases in rural population. Between 1980 and 1990, India's land per person decreased by 14 percent, whereas Pakistan's decreased by 22 percent. The land/people ratio has stayed essentially stable in Iran, despite a decreased rural population increase. In South Asia, there is essentially no undeveloped yet useable land. All the best land has already been taken, and what is left cannot be used for agriculture on a long-term basis.

Land tenure: Farmers will be hesitant to invest in land conservation measures if their future rights to use these resources are uncertain. Tenancy and open access resources; Farmers will be hesitant to engage in land conservation measures if their future rights to use these resources are uncertain. This predicament is caused by two types of property rights: tenancy and open access resources. Tenant occupation of farmland is still relatively common, despite many years of legislative and land reform efforts. The proprietor is typically a city dweller, and the property is farmed by tenants who pay a rental fee. The landlord-tenant relationship is frequently positive, and the tenant commonly stays on the same farm for many years. However, because they are only concerned with the immediate crop, such renters are not motivated to keep the land in excellent condition. Open access land resources are ones that everyone, including the impoverished and landless, can utilize without the need for an ongoing usufruct or tenure. This is especially true of forest lands that are ostensibly owned by the government but are populated on a squatter basis. The terms "common property" and "open access resources" are not interchangeable. The use of common property resources is limited to members of a community, village, or clan, and is subject to socially imposed limits. Pastoralists, for example, have norms regulating when areas must be rested from grazing, and villages regulate collective woods cutting. There are no such restrictions on free access land. With no legal foundation for its use, there is no motivation to farm the land for anything other than current requirements. This, along with water erosion, is a major source of deforestation.

***Economic pressures and attitudes*:** Farmers face enormous economic challenges as a result of their small land holdings, which make it difficult for them to obtain enough food and revenue to meet their immediate requirements. Because of the short-term pressures, labor, land, and capital resources cannot be diverted to land care, such as green manuring or soil conservation buildings. This is also the underlying cause of two other direct causes, incorrect crop rotations and unbalanced fertilizer use, which were mentioned earlier. A shift in economic sentiments is a contributory component that isn't often seen by outside observers. In the past, most farmers accepted their situation, even if it was one of relative poverty. Increased economic pressures have resulted from modern communications and influence, which have led to higher goals and, as a result, higher income requirements.

***Poverty*:** The region's countries have made significant economic success, including rises in gross domestic product per capita. It's debatable whether the rural poor's true well-being has improved as a result of these changes. Many farmers stay on or around the poverty line, which is defined as having access to basic requirements of life. Land degradation is a result of poverty. It is almost probable that affluent farmers keep their soils in better condition than poorer farmers.

***Population increase*:** The continued expansion in rural, agricultural, population, in addition to land scarcity, is the second major driver of degradation. For six countries, annual population growth rates range from 2.1 to 3.6 percent from 1980 to 1990. (For Afghanistan the figure is affected by migration and war). Only Sri Lanka, with a population growth rate of 1.4 percent, has achieved significant headway in reducing the rate of population expansion. The urban population is growing at a greater rate than the rural population. However, the movement toward urbanization is insufficient to alter the fact that absolute rural population levers have climbed and continue to increase. Of 1990, rural populations in Bangladesh, Bhutan, India, Nepal, and Pakistan were 17-32 percent greater than in 1980. In absolute terms, the magnitude is largest in India, where already highly populated rural areas had 79 million more people in 1990 than they did ten years before.

PROCESSES OF SOIL DEGRADATION

Soil degradation is classified in many ways. The Global Assessment of Soil Degradation (GLASOD) recognizes four major types of soil degradation:

- Erosion: water erosion, wind erosion,
- chemical deterioration (including organic matter decline, salinization, acidification, and pollution), and
- physical deterioration (such as compaction, sealing, waterlogging, and urbanization).

Erosion Degradation

Soil erosion is a natural occurrence that has an impact on all landforms. Soil erosion in agriculture refers to the natural physical forces of water and wind, as well as forces connected with farming activities such as tillage, eroding away a field's topsoil. Erosion encompasses three separate actions: soil detachment, movement, and deposition, whether caused by water, wind, or tillage. Topsoil, which is rich in organic matter, fertility, and soil life, is either moved "on-site" to build up over time or transported "off-site" to fill in drainage channels. Soil erosion affects agriculture productivity while also polluting nearby watercourses, marshes, and lakes. Soil erosion can be a gradual process that goes unnoticed, or it can happen quickly, resulting in significant topsoil loss. Other major soil degradation characteristics that might speed up the erosion process include soil compaction, low organic matter, loss of soil structure, poor internal drainage, salinization, and soil acidity issues.

Water Erosion

Water erosion has become the focus of soil conservation initiatives due to its ubiquitous occurrence and the severity of its on-site and off-site effects.

The following elements influence the rate and amount of soil erosion caused by water:

Rainfall and Runoff: The higher the risk for erosion, the greater the intensity and duration of a downpour. The action of rainfall on the soil surface can spread aggregate material and break down soil aggregates. Lighter aggregate materials, such as extremely fine sand, silt, clay, and organic matter, are easily removed by raindrop splash and runoff water; larger sand and gravel particles require more raindrop energy or runoff volumes. During short-duration, high-intensity thunderstorms, soil movement (raindrop splash) is usually highest and most obvious. Although the amount of soil loss generated by long-lasting and less-intense storms is not normally as dramatic or obvious as that caused by thunderstorms, it can be significant when compounded over time. When there is extra water on a slope that cannot be absorbed into the soil or is trapped on the surface, surface water runoff occurs. Increased runoff is caused by less infiltration due to soil compaction, crusting, or freezing. Agricultural land runoff is highest in the spring, when the soils are normally moist, the snow is melting, and the vegetation cover is limited.

Soil Erodibility: Soil erodibility is a measurement of a soil's ability to resist erosion based on its physical features. Texture is the most important factor in erodibility, but structure, organic matter, and permeability also play a role. Soils with faster infiltration rates, higher organic matter levels, and enhanced soil structure are more resistant to erosion in general. Silt, very fine sand, and certain clay-textured soils are less erodible than sand, sandy loam, and loam-textured soils. Increases in soil erodibility are caused by tillage and farming techniques that lower soil organic matter levels, induce poor soil structure, or result in soil compaction. Compacted subsurface soil layers, for example, can reduce infiltration and increase runoff. Infiltration is also reduced by the creation of a soil crust, which tends to "close" the surface. A soil crust may reduce the amount of soil loss from raindrop impact and splash on some locations, but a commensurate increase in runoff water can lead to more serious erosion concerns. The erodibility of a soil is also affected by past erosion. Because of their inferior structure and lower organic matter, many exposed subsurface soils on eroded sites are more erodible than the original soils. Lower nutrient levels in subsoils lead to lower crop yields and, in general, worse crop cover, resulting in less crop protection for the soil.

Slope Gradient and Length: The higher the risk of erosion, the steeper and longer the slope of a field. Because of the increased buildup of runoff, soil erosion by water increases as the slope length grows. Because of the increased velocity of water, which allows for a greater degree of scouring, combining small fields into bigger ones generally results in longer slope lengths with enhanced erosion potential (carrying capacity for sediment).

Cropping and Vegetation: If the soil has no or very little vegetative cover of plants and/or crop wastes, the risk of soil erosion increases. Plant and residue cover shields the soil from raindrop impact and splash, slows runoff water flow, and allows extra surface water to penetrate. The efficiency of plant and/or agricultural leftovers in controlling erosion is determined by the type, extent, and quantity of cover. Controlling soil erosion is most effective when vegetation and residue mixtures completely cover the soil and intercept all falling rains at or near the surface (e.g., forests, permanent grasses). Residues that have been partially integrated and residual roots are also essential because they provide routes for surface water to enter the soil. The efficiency of any protective cover is also determined by the amount of protection available at different times of the year in relation to the amount of erosive rainfall that occurs during these times. Crops that provide a full protective cover for a significant portion of the year (e.g., alfalfa or winter cover crops) can reduce erosion significantly more than crops that leave the soil bare for a longer period (e.g., row crops), especially during periods of high erosive rainfall, such as spring and summer. Crop management approaches that emphasize contour farming and strip-cropping techniques can help to prevent erosion even further. Leave a residual cover greater than 30% after harvest and over the winter months to reduce most erosion on annual row-crop land, or inter-seed a cover crop (e.g., red clover in wheat, oats after silage corn).

Tillage Practices: Tillage activities increase the possibility for soil erosion by water, depending on the depth, direction, and timing of plowing, the type of tillage equipment used, and the number of passes. In general, the less vegetation or residue cover at or near the surface is disturbed, the more successful tillage is at reducing water erosion. Water erosion can be reduced by using minimum till or no-till methods. Tillage and other procedures that are carried out up and down field slopes create channels for surface water runoff, which can hasten soil erosion. Cross-slope cultivation and contour farming practices reduce surface water runoff concentration and soil mobility.

Forms of Water Erosion

***Sheet Erosion*:** The movement of soil caused by raindrop splash and runoff water is known as sheet erosion. It usually happens in a straight line on a flat slope and goes unnoticed until the productive topsoil has been removed. The eroded dirt is deposited near the bottom of the slope or in low places. Other markers include lighter-colored soils on knolls, changes in soil horizon thickness, and low crop yields on shoulder slopes and knolls.

Figure 4: Sheet erosion

Source: Google Images

***Rill Erosion*:** Surface water runoff concentrates, generating narrow but well-defined channels, resulting in rill erosion.

Figure 5: Rill Erosion

Source: https://agbmps.osu.edu/scenario/inter-rill-and-rill-erosion

When the dirt has been washed away, these separate channels are called rills when they are small enough to not interfere with field machinery activities. Rills are filled in every year as part of tillage operations in many cases.

***Gully Erosion*:** Gully erosion is a more advanced kind of rill erosion in which surface channels are degraded to the point that they represent a tillage nuisance. Due to gully erosion, several farms in Ontario are losing significant amounts of topsoil and subsoil each year. The creation of gullies or the enlargement of existing gullies is mainly caused by faulty outlet design for local surface and subsurface drainage systems. Sloughing and slumping (caving-in) of bank slopes is caused by the soil instability of gully banks, which is frequently connected with groundwater seepage. Such failures are most common in the spring, when the soil water conditions are ideal for the problem. If remedial measures are not developed and built appropriately, gully development can be difficult to regulate. Control methods must consider the reason of the increased water flow across the landscape and be capable of diverting runoff to a correct exit. Gully erosion causes significant area to be taken out of production and creates dangerous conditions for agricultural tractor workers.

Figure 6: Gulley Erosion

Source: Google Images

***Bank Erosion*:** Surface water runoff and subsurface drainage systems are discharged by natural streams and manmade drainage channels. The gradual undercutting, scouring, and slumping of these drainageways is known as bank erosion. Bank erosion can be caused by poor construction practices, lack of upkeep, uncontrolled animal access, and crops too close together. Bank erosion is further exacerbated by poorly designed tile outlets. Some are inoperable because they lack a rigid outlet pipe, have an insufficient or non-existent splash pad, or have outlet pipes that have been damaged by erosion, machinery, or bank cave-ins. Loss of arable farmland, undermining of buildings such as bridges, increased need to clean and maintain drainage channels, and washing out of lanes, roads, and fence rows are all direct effects of bank erosion.

Figure 7: Bank Erosion

Source: Google Images

Effects of Water Erosion

On-Site: The consequences of water-induced soil erosion go beyond the loss of rich topsoil. The loss of natural nutrients and applied fertilizers has a direct impact on crop emergence, growth, and yield. Erosion can disrupt or eliminate seeds and plants. Organic matter from the soil, residues, and any applied manure are generally light and can be easily carried off the field, especially during the spring thaw. Pesticides could potentially be moved away from the site via eroding soil. The loss of soil can impact the quality, structure, stability, and texture of the soil. The removal of smaller particles or entire layers of soil or organic matter, as well as the breakup of aggregates, can weaken the structure and potentially affect the texture. Textural changes can decrease the soil's water-holding capacity, making it more vulnerable to harsh weather like drought.

Off-Site: The off-site effects of water-induced soil erosion are not necessarily as obvious as the on-site ones. Eroded soil deposited down slope prevents or delays seed emergence, buries young seedlings, and demands replanting in impacted regions. Sediment can also accumulate on down-slope homes, causing road damage. Sediment in streams and watercourses can hasten bank erosion, clog stream and drainage channels, fill reservoirs, harm fish habitat, and affect downstream water quality. Pesticides and fertilizers affect or pollute downstream water sources, marshes, and lakes, as they are frequently transferred with eroding soil. Controlling "non-point" pollution from agricultural land is an essential topic due to the potential severity of some of the off-site effects.

Wind Erosion

Wind erosion occurs primarily in sandy and organic or muck soils in vulnerable places. It has the potential to cause significant soil and property losses under the correct circumstances. Suspension, saltation, and surface creep are three ways for soil particles to move, depending on their size and wind strength. The following elements influence the rate and degree of wind-driven soil erosion:

***Soil Erodibility*:** The wind carries very fine soil particles high into the air and transports them over long distances (suspension). Fine-to-medium sized soil particles are raised into the air and then fall back to the soil surface, causing damage to crops and dislodging more soil (saltation). Wind dislodges larger soil particles that are too massive to be lifted from the ground and rolls them along the soil surface (surface creep). The abrasion caused by windblown particles breaks down stable surface aggregates, increasing soil erodibility even more.

***Soil Surface Roughness*:** Surfaces that are not rough provide low wind resistance. In a windstorm, however, ridges created by tillage might dry up faster, resulting in more loose, dry soil accessible to blow. Surfaces of soil fill in with time, and abrasion breaks down the roughness. As a result, the surface is smoother and more vulnerable to wind. Excess tillage can contribute to erosion and soil structure disintegration.

***Climate*:** The amount of soil erosion is proportional to the speed and duration of the wind. When soil moisture levels are very low at the surface of overly drained soils or during droughts, particles are released for wind transport. This effect can also be seen when the soil surface freezes over during the winter months. Wind erosion is indicated by the accumulation of soil on the leeward side of barriers such as fence rows, trees, or buildings, or by brown snow cover during the winter.

Unsheltered Distance: The absence of windbreaks (trees, shrubs, agricultural leftovers, etc.) causes the wind to move soil particles across longer distances, increasing abrasion and soil erosion. The most exposed areas are knolls and hilltops, which suffer the most.

Vegetative Cover: Wind erosion is exacerbated in some areas due to a lack of permanent plant cover. The most susceptible soil is loose, dry, bare soil; yet, crops that create little residue (such as soybeans and many vegetable crops) may not provide enough resilience. Even crops that produce a lot of leftovers may not be able to preserve the soil in extreme instances. A cover crop with a suitable network of living windbreaks, in combination with good tillage, residue management, and crop selection, is the most effective protective vegetative cover.

Effects of Wind Erosion

Sandblasting of young seedlings or transplants, burial of plants or seed, and seed exposure are all examples of how wind erosion harms crops. Crops are damaged, causing costly delays and necessitating reseeding. Plants injured by sandblasting are more susceptible to disease, resulting in lower yields, worse quality, and lower market value. Wind erosion can also wreak havoc on operations, hindering timely field activities. Soil drifting is a fertility-depleting process that can result in poor crop growth and yield losses in fields where wind erosion is a common occurrence. A textural alteration in the soil is caused by continuous drifting of a region. The loss of fine sand, silt, clay, and organic particles from sandy soils reduces the soil's ability to store moisture. This exacerbates the situation by increasing the soil's erodibility. Wind-blown dirt removal from fence rows, created drainage channels and roads, and around structures is a time-consuming and expensive process. Furthermore, soil nutrients and surface-applied pollutants might be conveyed with the soil particles, resulting in off-site effects. Blowing dust can also be harmful to people's health and pose a threat to public safety.

Tillage Erosion

Tillage erosion is the process of soil redistribution caused by tillage and gravity. It causes progressive down-slope movement of soil, resulting in substantial soil loss on upper slopes and buildup on lower slopes. This type of erosion is a common way for water to get into the ground. Soil is moved to convergent sections of a field where surface water runoff collects through tillage. Furthermore, uncovered subsoil is particularly erodible by water and wind. Tillage erosion has the greatest potential for "on-site" soil movement and, in some situations, can cause more erosion than wind or water. The following factors influence the rate and magnitude of soil erosion caused by tillage:

- ***Type of Tillage Equipment*:** Lifting and carrying tillage equipment will move more soil. A chisel plow, for example, leaves significantly more crop residue on the soil surface than a moldboard plow, but it may move as much soil as the moldboard plow and move it further. Tillage erosion can be reduced by employing implements that do not move a lot of soil.
- ***Direction*:** Depending on the direction of tillage, tillage instruments like a plow or disc push soil up or down slope. When tilling in the down-slope direction, more dirt is typically moved than when tilling in the up-slope direction.
- ***Speed and Depth*:** The amount of soil moved is influenced by the speed and depth of tillage operations. Deep tillage disturbs more soil, whereas faster tillage moves the dirt further.
- ***Number of Passes*:** The number of passes made by tillage equipment is reduced, which lowers soil movement. It also leaves more crop residue on the soil surface and minimizes soil aggregate pulverization, both of which help to withstand water and wind erosion.

Effects of Tillage Erosion

Crop development and yield are harmed by tillage erosion. Due to weak soil structure and loss of organic matter, crop growth on shoulder slopes and knolls is slow and stunted, and it is more subject to stress under harsh conditions. Changes in soil structure and texture can make the soil more erodible, exposing it to more erosion from the pressures of water and wind. Tillage erosion can include the shifting of subterranean soil in extreme circumstances. Subsoil that has been shifted from upper-slope to lower-slope locations has the potential to bury the productive topsoil in lower-slope areas, affecting crop development and productivity. Tillage-eroded fields have demonstrated soil loss of up to 2 m in depth on upper-slope sites, as well as yield decreases of up to 40% in corn, according to research. In extreme circumstances, remediation entails relocating displaced soils to higher-slope locations.

Chemical Soil Degradation

Chemical degradation of soil refers to unfavorable changes in soil chemical behavior as a result of human involvement, resulting in a decrease in soil quality. Chemical degradation affects around 240 million hectares of land, with 229 million hectares lost to nutrient loss (136 million hectares), salinization (77 million hectares), and acidity (6 M ha). Soil contamination affects another 11 million hectares. About 12% of the overall area affected by human-induced soil deterioration is due to chemical soil contamination (Oldeman 1992). Around 12% of degraded soils in Africa, Asia, Central America, and Europe are chemically impacted. Chemical degradation accounts for around 30% of deteriorated soils in South America. Chemical soil deterioration is not a major concern in North America or Oceania. In Africa and South America, nutrient loss is the most common type of chemical soil degradation, but in Asia, salinization is the most common type of chemical soil degradation. Agricultural mismanagement (56%) and deforestation (28%) are the main causes of chemical soil deterioration (Oldeman 1992).

Table 10: Global extent of soil degradation (M ha)

Regions	Loss of nutrients	Salinization	Acidification	Total
Africa	65	15	1	62
Asia	15	53	4	72
South America	68	2	–	70
Central America	4	2	–	6
North America	–	+	+	+
Europe	3	4	+	26
Oceania	+	1	–	1
World	136	77	6	229

Source: Data from Oldeman (1992)

Loss of Nutrients and/or Organic Matter

Although it is commonly associated with low-input agriculture, nutrient loss occurs in a variety of other contexts. It can happen if agriculture is done on low or moderately fertile soils with insufficient manure and fertilizer application. The clearance of natural vegetation causes a rapid loss of organic matter and nutrients in the soils. Other significant sources of nutrient and organic matter loss include biomass burning, shifting agriculture, and erosion. Human-induced soil nutrient depletion refers to the process of the soil nutrient stock reducing as a result of continual nutrient mining without adequate replacement of nutrients gathered in agricultural products, as well as nutrient losses owing to accelerated soil erosion and leaching (Tan et al. 2005). Human efforts for the use of vegetation and soil resources enhance natural erosion and leaching. A negative balance of nutrients and organic matter output (through harvesting, burning, leaching, etc.) and input (via manure and fertilizers, returning crop leftovers, flooding, etc.) is the most common cause of fertility depletion. As an example, the nutritional balance situations in several African countries are discussed below. In various African countries, the International Fertilizer Development Center (IFDC) conducted soil nutrient balance research. The methodological approach to estimating nutrient balances, depletion rates, and requirements integrates data on agricultural productivity, soil properties, and biophysical restrictions with estimation methods and procedures (Henao and Baanante1999). Land use, population-supporting capacity of land, crop productivity, and mineral and organic fertilizer use were among the facts and data connected to agricultural production. The method generates data for analysis and monitoring by combining attribute and geographic database systems with empirical and mechanistic models. The method draws on prior research on nutrient balances (Smaling et al. 1993). Due to the expansion of land use for agricultural production without effective application of external inputs, soil nutrient depletion is connected to food insecurity in emerging and least developed nations (Henao and Baanante 1999). Agricultural sustainability in these areas is jeopardized by the lack of fertilizer replenishment of nutrient-depleted soils (Sheldrick et al. 2002). Long-term fall in crop yields in many countries of Africa, Asia, and Latin America due to insufficient input and imbalanced fertilization (FAO/UNDP/UNEP/World Bank 1997). The nutrient budget approach was utilized by Smaling et al. (1993) to emphasize the critical condition of soil nutrient depletion in several African countries. In Asia, soil nutrient depletion and imbalance are also common problems. Dobermann et al. (1995) calculated nutritional balances for rice-based farming systems in seven eastern Asian nations and found a K shortfall of 25 to 70 kg ha1 year1. Lin et al. (1996) calculated nutrient balances on 71 rice paddy farms in South China and discovered that N and P were in abundance, but K was in short supply. Negative N, P, and K balances were reported by Mutert (1996) in Bangladesh, Indonesia, Myanmar, the Philippines,

Thailand, and Vietnam. Nutrient loss can occur as a result of:

Leaching of Nutrients: The downward movement of soil elements in solution and suspension with percolating water is referred to as leaching. It distributes and removes large amounts of materials if there is a positive water balance, that is, if there is enough evapotranspiration water excess to allow for effective downward movement. In humid climates, leaching is a significant nutrient removal mechanism. Even in natural vegetation, leaching occurs, but agricultural activities can dramatically increase leaching losses (Havlin et al. 1999). Nutrient leaching is more likely in soils with high water infiltration rates and poor nutrient retention capacity, such as sandy soils and well-structured ferralitic soils (Oxisols, Ultisols) with low activity clays and low organic matter contents. The mobility of nutrients in soil varies; nitrates are more mobile than other anions. The negatively charged matrix of the top soil does not attract nitrate ions. The nitrification of ammonia received from fertilizers and soil organic matter produces nitrates in the soil on a constant basis. In sandy and low-retentive soils, nitrate leaching is a severe problem (Robertson 1989). It frequently contaminates groundwater and open reservoir water in agricultural environments. Sulfates and phosphates are both quickly leached from agricultural soils, but sulfates are less mobile. Phosphates are immobilized by soil colloids, calcium, aluminum, iron, and manganese. The electrically neutral soil solution that percolates through the soil profile comprises anions and comparable numbers of cations. As a result, cations are also leached, with calcium and magnesium being more sensitive than potassium (Pieri 1989). After applying potassium chloride or potassium sulfate fertilizers to sandy soils, significant amounts of magnesium can be leached (Havlin et al. 1999). Calcium and magnesium are leached in significantly greater amounts than potassium.

Residue Burning: When biomass residues, crop residues, or litter are burned, most of the nitrogen (N) and sulfur (S) is lost, but mineral nutrients like phosphorus (P) and potassium (K) are retained. Wheat, oat, and flax waste were burned in an open container in a Manitoba laboratory (Heard et al. 2006). The remaining ash was collected and weighed after the burn. The mineral elements (P and K) remained in the ash after much of the N and S was oxidized and dissipated as volatile gases. N losses were 98–100%, S losses were 75%, P losses were 21%, and K losses were 35%, according to the research. The P and K losses were determined to be the result of smoke and ash escaping from the burn container. Depending on wind and other environmental circumstances, some of this ash may be redeposited onto the field after a field fire. Other mineral nutrient losses are likely to be like those of P and K. After a fire, the weather conditions often play a significant impact in nutrient loss from the crop. High winds, for example, might sweep ash from the area or pile it into drifts. Biomass leftovers are burned by shifting cultivators to provide nutrients to their crops. Such nutrients are only beneficial for a short time since severe rains following soil preparation often wash away the bases produced by fire. Furthermore, when the residue cover is lost and the soil is exposed, soil erosion from both wind and water is more likely. The average yearly soil organic matter reductions were 0.03 and 0.07 percent in a western Canada research where cereal residue was burned annually for 19 years (Biederbeck et al. 1980).

Erosion: Soil type, cover, and fertility; slope of the terrain; amount, manner, and time of fertilizer application; intensity, amount, and timing of rainfall (Gascho et al. 1998); and soil management methods all influence nutrient losses by water erosion (Mello 2002). In soil management strategies that conserve plant leftovers, nutrient losses are predicted to be decreased. Plant leftovers, however, can be washed away under such conditions (Burwell et al. 1975), becoming sources of soluble nutrients (Johnson et al. 1979), which can be lost to water erosion. Water erosion causes significant losses of calcium and magnesium (Schick et al. 2000). This happens because these elements are strongly absorbed by colloids in the soil, making them simple to travel with the sediment, and they also occur in large concentrations in erosion water. In slashed and burned sites in Bangladesh's southeastern hills, Gafur et al. (2000) observed losses of 61 kg Ca, 13 kg Mg, 13 kg K, 0.14 kg P, 0.20 kg S, 0.05 kg Cu, 6.7 kg Fe, 6.1 kg Mn, and 0.065 kg Zn owing to erosion.

***Crop Removal of Nutrients*:** Soil and climate variables influence crop nutrient loss. Plant nutrient uptake is influenced by soil variables such as accessible soil nutrient supply, moisture content, aeration, compaction, soil temperatures, pH, nutrient imbalances, and other factors. The values of nutrients eliminated by harvested crops are listed in Table 11. These results might be multiplied by a factor of 2–3 if nutritional residues were included. As a result, crops alone lose a significant quantity of soil nutrient, which must be replaced with fertilizers. If residues are removed from the field as well, the loss increases dramatically. Harvested crops lose 50–100 kg ha1 N, 10–20 kg ha1 P, and 15–30 kg ha1 K from field crops on average. Forage crops remove 100–200 kg potassium per hectare.

Acidification

The concentration of hydrogen (H+) ions in soil solution indicates the acidity of the soil. The soil is said to be acidic when (H+) ions prevail over (OH) ions. Soil pH reflects the relative abundance of (H+) and (OH) ions. The pH scale goes from 1 to 14 and is based on the negative logarithm of hydrogen concentration (moles per liter). The neutral pH value is 7.0, with values below 7.0 being acidic and over 7.0 being alkaline. Because the pH scale is logarithmic, soil with a pH of 5 is 10 times more acidic than soil with a pH of 6, and soil with a pH of 7 is 100 times more acidic. There are two forms of soil acidity: active acidity, which is caused by the presence of (H+) ions in solution, and reserve or exchangeable acidity, which is caused by the presence of exchangeable H+ and Al3+ ions on soil colloids' exchange sites. The lower the pH and the higher the acidity of the soil, the greater the percentage of exchange sites occupied by aluminum and hydrogen.

Table 11: Aluminum sensitivity/tolerance of some plants

Categories	Plants
Highly sensitive	Durham wheat, barley, lentils, chickpeas, alfalfa, strawberry, berseem, buffel grass, tall wheatgrass
Sensitive	Canola, red clover, balansa clover, white clover
Tolerant	Whistler, diamond bird wheat, ryegrass, tall fescue, subterranean clover, chicory
Highly tolerant	Narrow leaf lupins, oats, triticale, cereal rye, cocksfoot, paspalum, yellow and slender serradella, Consol love grass

Source: Osman, K. T. (2014).

Acidification can occur naturally in soils made from acidic parent materials due to the release of acid-forming chemical compounds, in high-rainfall areas due to prolonged leaching of bases, at higher elevations due to erosion, and under prolonged weathering due to the predominance of iron and aluminum oxides. However, agricultural operations have the potential to accelerate or cause soil acidity in a short period of time. Acidification in agricultural soils can occur as a result of nitrogen fertilizer application, nitrate leaching, produce removal, and soil organic matter building (Upjohn et al. 2005). The following are some of the reasons of acidification:

Fertilizers: The quantity of acidity caused by nitrogenous fertilizers varies depending on the type of fertilizer used. Fertilizers containing nitrogen in the form of ammonium, such as ammonium sulfate, acidify soil within weeks of application. The microbial oxidation of ammoniacal fertilizers, which might be strong bases themselves, is the most important acid-forming reaction for fertilizers.

- NH3 +2O2=H++ NO3− +H2O (nitrification of ammonia)
- NH4NO3+2O2=2H++2NO3−+ H2O (nitrification of ammonium nitrate)
- CO(NH2)2+4O2= 2H+ +2NO3–+H2O+ CO2 (hydrolysis of urea and nitrification of products)

The compounds listed above are the most prevalent synthetic N inputs to agroecosystems, and will oxidize to nitric acid under the well-drained, aerobic conditions of typical dryland agricultural soils. Unless all the nitrate is leached out of the soil, calcium nitrate and sodium nitrate have a neutralizing impact on soil acidity, but they are expensive and only used in horticulture. Superphosphate fertilizer is not directly acidifying the soil when used on crops and pastures. Soil acidification can be achieved by using pure sulfur or "flowers of sulfur." Dr. Lloyd A. Peterson, of the Arlington Agricultural Research Station in South Central Wisconsin, devised a long-term experiment known as the 444 NPK fertility study in 1962. Soil samples were tested and reported by Barak et al. after 30 years of fertilizer inputs (1997). At the long-term fertility trial in Arlington, Wisconsin, ammonium nitrate and urea N fertilizers produced soil acidity. For 0–150 lb N/acre yearly treatments, mean pH values ranged from 5.6 to 4.8 in water and from 5.5 to 4.1 in 1 M KCl. Below is a list of fertilizers with different levels of acidifying potential.

- ***Slightly acidifying fertilizers***: Solutions of urea, ammonium nitrate, and urea ammonium nitrate. Because they contain ammonium or create ammonium when applied to the soil, these goods are mildly acidifying. To neutralize the acidity, 1.8 kg of pure calcium carbonate is required for every kilogram of real nitrogen injected.
- ***Moderately acidifying fertilizers***: Phosphate of diammonium (DAP). Diammonium phosphate is the most common kind of phosphorus fertilizer in many nations, and it has a mild acidifying impact when used. Complete neutralization requires 5.25 kg of limestone for every kilogram of N added as DAP. If phosphorus fertilizer is needed, triple superphosphate (0-46-0) can be used because it has no effect on the pH of the soil.
- ***Severely acidifying fertilizers***: Monoammonium phosphate,

ammonium sulfate (MAP). These fertilizers are extremely acidifying and should be avoided at all costs. To offset the effects of 1 kg of these fertilizers, approximately 8 kg of agricultural limestone is required. If a planter sprayed 50 kg of MAP or ammonium sulfate, for example, 400 kg of lime would be required. Farmers typically use fertilizers but not lime. As a result, the soil becomes acidic over time. The amounts of lime needed for a certain amount of fertilizer are provided in the table below.

Table 12: Lime requirement for different fertilizers with acidifying effect

Nitrogen source	Lime requirement, kg pure CaCO3 per kg N applied (Glendinning 2000)
Ammonium sulfate	5.2
Anhydrous ammonia	1.8
Ammonium nitrate	1.8
Urea	1.8
MAP	5.0
DAP	3.1

Source: Osman, (2014).

***Leaching of Nitrate*:** Nitrate leaching is a major contributor to agricultural soil acidity. Nitrate is made from the biological oxidation of ammonium or is added to fertilizers that include nitrate. Ammonium is derived from fertilizers and the mineralization of organic matter in the soil. The chemical reactions that produce nitrate nitrogen from ammonium mildly acidify the soil. Plants counteract the acidity by releasing an alkaline material as they absorb nitrate nitrogen. The acid/alkali balance of the soil surrounding the roots remains in balance as the plants continue to absorb all the nitrate nitrogen. Increased soil acidity occurs from leaching, which disrupts the acid/alkali balance. The effects of nitrate leaching on the ecosystem are significant. Nitrates leached from agricultural fields have the potential to damage ground and surface water.

***Removal of Produce*:** Grain, pasture, and animal products are slightly alkaline, and removing them from the soil over time will lower the pH. When only a small amount of produce is eliminated, as in wool production, the system remains almost balanced. When a substantial amount of produce is removed, such as when creating hay (especially clover or alfalfa hay), the soil becomes significantly more acidic. Burning product, such as stubble, does not modify the soil's acid/alkali balance; instead, it causes a redistribution, leaving alkali near the soil surface as ash. If the ash is washed away as a result of rains following a fire, the soil will become more acidic.

***Buildup of Soil Organic Matter*:** Increased organic matter has various advantages, including improved soil structure, but it also raises soil acidity. Acidification induced by a buildup of organic matter is not permanent, and it can be reversed if the organic matter decomposes. If the topsoil containing the organic matter is eroded or destroyed, however, the soil's acidity will be permanently altered.

***Acid Rain*:** Acid deposition causes depletion of nutrient-base cations (Ca, Mg, K, and others) in soils, according to Watmough and Dillon (2003). Several forested places in North America have experienced acidification and nutritional cation depletion as a result of acid deposition. Continued regional nitrogen and sulfur imports are a source of worry due to base cation leaching, increased soil Al availability, and the accumulation and eventual transmission of acidity from forest soils to streams (Fenn et al. 2006). Acid rain is the result of acid deposition. It occurs when acidic materials such as sulfur dioxide (SO2), ammonia (NH3), nitrogen oxides (NOx), and acidic particles emitted from fossil fuel combustion and other industrial processes undergo complex chemical reactions in the atmosphere and fall to the earth as wet deposition (rain, snow, cloud, fog) or dry deposition (dust) (dry particles, gas). Acid rain's impact on the soil is mostly determined by the soil's ability to neutralize the acid. Calcareous soils are the most successful in neutralizing acidity in acid rains and preventing the pH of the soil from dropping. High soil solution concentrations of SO42 and NO3 are linked to anthropogenic soil acidification caused by acid rain (Matzner and Davis 1996).

Effects of Soil Acidity

Some or all the following issues can occur in acidic soils:

***Nutrient Availability*:** The accessible nutrient forms in soil are soluble and exchangeable ions. The solubility of elements and compounds in soil is regulated by soil pH, which also controls the availability of plant nutrients. When the pH of the soil falls below 6.5, or when the soil becomes more acidic, the solubility of some elements, such as Fe, Al, Mn, Cu, and Zn, increases while the solubility of others, such as Ca, Mg, and Mo, decreases. When soil pH rises above 7.0, Ca, Mg, and Mo solubility increases, while Fe, Al, Mn, Cu, and Zn solubility decreases. Although aluminum is abundant in soil, it is not a necessary element for plants. Al, Fe, and Mn solubility increases to the point where they become poisonous to many plants in very acidic soils (pH < 5). In addition, acidic soils can cause Ca and Mg deficiency in plants. Phosphorus precipitates with Al, Fe, and Mn as polyphosphates at low pH (5.5). P precipitates with Ca at high pH (>8.0). P availability is lowered to deficient levels when soil acidity and alkalinity are both high. In the pH range of 6.5 to 7.0, P availability is usually higher. Strongly acidic and strongly alkaline soils can also cause boron shortage. Low pH limits the availability of macronutrients (Ca, Mg, K, P, N, S), as well as Mo and B. At low pH, however, the availability of most micronutrients (Fe, Mn, Zn, Cu, and Co) increases. Plant nutrition levels are best when the pH is between 5.5 and 6.5.

***Toxicity of Iron, Aluminum, and Manganese*:** When the pH of the soil is 4.8 or lower, nutrient poisoning might ensue (Slattery et al. 1999). Aluminum (Al) and manganese (Mn) are the two most hazardous elements in acidic soils (Slattery et al. 1999). Aluminum and manganese become more accessible in the soil solution in strongly acidic soils (pHW4.3) and are damaging to plant roots. On acidic soils, aluminum toxicity is the most common plant symptom, resulting in root stunting (Slattery et al. 2000). Reduced root growth obstructs nutrient and water intake, resulting in lower output. Some plants are more resistant to high levels of Al in the soil solution than others. In acidic soils, important productive plants including alfalfa, phalaris, canola, and barley are difficult to establish and flourish. Low pH and toxic aluminum (Yokota and Ojima 1995) have irreversible effects on alfalfa establishment. Deep-rooted perennial pastures (such as alfalfa and phalaris) are being grown to halt the acidification process. Aluminum may obstruct Ca uptake and cause P to precipitate in the plant body. It has the potential to disrupt P metabolism, as well as ATP and DNA activities. Plant roots become stunted and branchless, with brown root tips and lateral roots. Aluminum toxicity causes symptoms in the leaves that are like P insufficiency. In soils with a pH below 4.0, oxidized Fe toxicity is common. In soils with 200–5,000 mg kg1 Mn, manganese poisoning is likely. In solution culture studies of cotton (Gossypium hirsutum L.), sweet potato (Ipomoea batatas L.) (Mortley 1993), sorghum (Sorghum bicolor) (Mgema and Clark 1995), and wheat (Triticum aestivum L.), manganese concentrations in the range of 0.2–12 mM have been reported to cause severe growth limitations (Taylor et al. 1991).

Table 13: Critical levels of manganese for some plants

Manganese tolerance category	Plants	Critical leaf Mn level mg kg−1
Highly sensitive	Alfalfa, pigeon pea, barrel, and bar medics	200–400
Sensitive	White clover, strawberry clover, chickpea, canola	400–700
Tolerant	Sub-clover, cotton, cowpea, soybean, wheat, barley, triticale, oats	700–1,000
Highly tolerant	Rice, sugar cane, tobacco, sunflower, most pasture grasses, oats, triticale, tiga, currency, cereal rye	>1,000

Source: Upjohn et al. 2005

***Plant Growth*:** Most terrestrial plants require a slightly acidic soil (pH approximately 6.5) to thrive. Soil microorganisms are most active at this pH, and plant nutrients are plentiful. This delicate equilibrium is disrupted by excessive alkaline and acid conditions, and plant nutrients that were previously abundant can become insufficient or hazardous to plant growth. If the soil pH becomes excessively acidic, several critical nutrients like as phosphorus, calcium, magnesium, and molybdenum become unavailable. Plant output will be reduced in farming systems, resulting in lower profitability and a greater need on fertilizers to maintain any type of profitable agriculture. Plants prefer certain pH levels.

Table 14: Some plants and their suitable pH

pH range	Suitable crops
4.5–5.5	Azalea, Camellia, Rhododendron
5.5–6.0	Blackberry, chicory, gooseberry, magnolia, pineapple, potato, oats, rosemary
5.5–6.5	Apple, avocado, barley, cranberry, capsicum, melon, turnip, strawberry, wheat
6.0–6.5	Broad bean, bean, carrots, lemon, lentil, olives, parsley, pear, pumpkin, raspberry, soybean, squash, watermelon, white clover

6.0–7.5	Asparagus, broad bean, broccoli, cabbage, cauliflower, celery, cherry, chrysanthemum, cucumber, dahlia, French bean, garlic, grape, horse radish, lettuce, mulberry, mustard, onions, peach, pear, peas, peppers, plum, radish, spinach, sweet corn, tomato, turnips
6.5–7.5	Alfalfa, asparagus, avocado, barley, sugar beet, beetroot, grape vine, mushroom, spinach

Source: Jett (2005)

***Microorganisms*:** Rhizobium survival and persistence are limited by soil acidity. Aluminum levels beyond a certain threshold are harmful to Rhizobia, inhibiting nodule formation and the nitrogen fixation process (Slattery et al. 2001). Many Victorian soils have been discovered to have a pH range that inhibits clover nodulation. It has a substantial effect on pasture production and subsequent crop systems that rely on biological nitrogen input. Aluminum can have a metagenetic effect on Rhizobia when the soil pH is very low and the aluminum solubility is very high. These data imply that changing soil pH may affect microbial biodiversity over time. Nutrient toxicities and poor nutrient availability restrict bacterial development in heavily weathered acidic soil. Acidity alters the microbial ecology of the soil and reduces root and rhizosphere impacts. It slows the decomposition of organic matter and nutrient cycling (Siqueira and Morera 1997). The activity of nitrifying bacteria, which break down organic matter into ammonium and nitrate for subsequent plant absorption, is reduced by acidity. Fungi have a pH range that is lower than bacteria.

***Soil Fauna*:** The delicate equilibrium between groups of living organisms in the soil is influenced by soil acidity. In general, the soil fauna's ability to deal with large variations in soil pH is significantly diminished. In acidic soil conditions, most macrofauna, especially deep-burrowing species like worms and termites, decline in number, with most activity confined to the litter layer, where the pH is substantially higher and usually alkaline. Earthworm abundance has been linked to acidic soil conditions, according to Mele and Carter (1999). Some worm species can endure acidic soil conditions and contribute to litter decomposition, but they are also greatly influenced by the moisture conditions in their environment. Low soil pH levels are toxic to deep-burrowing and soil-eating worm species (Lavelle and Spain 1995).

Soil Salinization

If the electrical conductivity of the saturation extract (ECe) (saturating the soil with water and sucking the extract) is greater than 4 dS m1, the soil is saline (decisiemens per meter). Excess soluble salts, primarily chlorides and sulfates, as well as carbonates and bicarbonates of sodium, potassium, calcium, and magnesium, are found in saline soils. Except for halophytes, soil salinity is damaging to plants; it produces water stress through osmotic disturbances in plant tissue and toxicity of specific salt components. Soil salinity has a significant impact on seed germination, plant growth, and agricultural yield. It can sometimes result in crop failure. Some soils are saline by nature. Primary salinization or natural salinization are the methods that create them. Mismanagement of soil and crop, notably poor irrigation and drainage, which alters the hydrologic equilibrium, causes some soils to become saline. This is referred to as human-induced salinization or secondary salinization. Natural and human-induced salinization can occur in two climatic settings: desert and semiarid regions, and humid regions. Water shortage in arid and semiarid environments, caused by low rainfall and excessive evaporation, prevents critical salt leaching. A net capillary rise of water also occurs, bringing salts to the surface soil. Excess irrigation or poor drainage in humid locations, on the other hand, causes the groundwater table to rise to the root zone of plants, making the soil saline. Oldeman et al. (1991) calculated that human-induced salinization affects 76.6 million hectares of land worldwide, but they did not distinguish between irrigated and non-irrigated rainfed areas. Salinity is affecting broad areas of India, China, Pakistan, and Central Asian countries, according to a survey conducted by Ghassemi et al. (1995). This is due to excessive irrigation and a lack of drainage facilities. Due to the usage of saline water for irrigation, some soils have become saline. There are two key techniques to managing soil salinity:

- selection of salt-tolerant crops and
- removal of excess salts from the root zone.

There are a variety of salt-tolerant crops that can be produced. In other cases, removing salts through irrigation and drainage, as well as safely extracting salty water, may be monetarily unfeasible. The cost-effectiveness of reclamation is determined by salinity, water and labor costs, crop return, and other factors. If the soil is saline alone, reclamation is reasonably simple; however, if the soil is saline – sodic or sodic – it is more challenging.

Table 15: Global extent of human-induced salinization in(106Ha)

Continent	Light	Moderate	Strong	Extreme	Total
Africa	4.7	7.7	2.4	–	14.8
Asia	26.8	8.5	17.0	0.4	52.7
South America	1.8	0.3	–	–	2.1
North and Central America	0.3	1.5	0.5	–	2.3
Europe	1.0	2.3	0.5	–	3.8
Australia	–	0.5	–	0.4	0.9
Total	34.6	20.8	20.4	0.8	76.6

Source: Oldeman et al. (1991)

Physical Degradation of Soil

The breakdown of soil structure, dispersion of soil particles, closure of pores, compression and increased density, consolidation, compaction, and limited root penetration, poor infiltration, waterlogging and runoff, and faster erosion are all examples of physical deterioration of soil. These processes result in desertification in arid and semiarid areas, in addition to denudation. The end outcome is a reduction in ecosystem sustaining capacity. Globally, there are approximately 68.3 million hectares of compacted soil, accounting for 4% of human-induced soil deterioration. Soil compaction is a worldwide issue that is linked to mechanized agriculture. It has resulted in yield losses of 25–50% in several European and North American regions, as well as 40–90% in West African countries. Reduced, conservation, and timely tillage, large additions of organic matter, mulching, crop rotation, subsoiling, and controlled trafficking are all recommended for the long-term usage of compacted soils. Desertification is a land degradation process that occurs primarily in arid and semiarid regions as a result of human activity on ecosystems paired with adverse climate circumstances. According to the United Nations, desertification affects 70 percent of arid lands, or around 30 percent of the world's cultivable territory. Texture, structure, bulk density, porosity, soil water, air, and temperature are all important physical features of soils that affect soil fertility, productivity, and quality. Only soil texture is a permanent feature that cannot be easily changed among these. Soil management strategies can easily modify the other attributes. Tillage and manure addition can change the structure of the soil. Tillage activities break up clods and peds and change the size and quantity of large and small pores in the soil. The bulk density, water holding capacity, and aeration status are all affected as a result. Tillage activities frequently degrade soil structure, reduce porosity, and compact the soil. As a result of poor soil management, physical deterioration of the soil may occur. The following processes have been discovered to be responsible for much of the physical deterioration of soil.

- **Coalescence:** Fine soil particles are gradually deposited between individual aggregates, fusing them together to form a huge structure. Coalescence is especially common in soils that have been tilled in a very dry state.
- **Slaking:** When dry soil aggregates are wetted rapidly by rain or irrigation, structural collapse occurs. Soils with low organic carbon (less than 2%) are prone to slaking.
- **Dispersion:** Due to large forces of repulsion between clay particles, the slaking products disintegrate into individual particles. High quantities of exchangeable sodium produce these high forces.
- **Consolidation:** Destruction of pore space, primarily big pores, reduces soil volume. Soils with minimal organic carbon (less than 2%) are especially susceptible to compaction, although soils with rich organic matter have elastic elasticity and can rebound.
- **Aggregate Pulverization:** This occurs when too-dry soils are tilled, destroying aggregates, and turning them into fine powder. Tilling when the soil is excessively wet, on the other hand, breaks aggregates by smearing because soil strength is low. Tillage should preferably take place when the soil moisture level reaches the "lower plastic limit."

One of the most common types of soil degradation is physical deterioration. Surface sealing, surface crusting, hard-setting, and compaction (Pc); waterlogging (Pw); lowering of the water table (Pa); and organic soil sinking (Pa) (Ps). Desertification is a physical degradation process that has become a major source of worry in arid and semiarid areas.

Surface Sealing, Crusting, Hard-setting, and Compaction (Pc)

Surface Sealing

The term "soil sealing" refers to a change in the nature of the soil (typically surface soil) that causes it to become impermeable. The collapse of soil aggregates due to physical stresses, such as raindrop impact, or a chemical agent, such as excess exchangeable sodium, disperses soil particles. Improper tillage activities can also cause structural degradation. When soils are tilled in dry conditions, for example, peds are ground to fine powders. Infiltration and percolation are reduced when fine particles enter macropores. Tilling in damp conditions, on the other hand, causes the soil to pool. When damp, unsteady peds may sag. As a result, water and plant roots are unable to penetrate the surface soil. Many uses, such as agriculture and forestry, lose sealed soils, while biological soil services are significantly harmed or even eliminated. Changes in water flow patterns or habitat fragmentation may also have an impact on neighboring soils. According to current research, soil sealing is largely irreversible. Surface sealing may have a significant impact on soil hydrology. A surface seal, for example, will significantly restrict water movement into the soil (Ahuja 1983). Seal formation reduces porosity, pore size, and pore continuity while increasing soil density. A seal, according to McIntyre (1958), is made up of two parts:

- an upper skin zone of roughly 0.1 mm due to raindrop impact compaction and
- a washed-in zone of decreased porosity of around 1.5 mm due to particle migration into the soil with water

According to Agassi et al. (1981), there are two complementary mechanisms for seal formation:

- Wetting and raindrop impact create a physical degradation of soil aggregates, and
- Clay particles migrate into the soil with infiltrating water, clogging pores, and forming a washed-in layer of low permeability.

Reduced overall porosity raises bulk density (Assouline, 2006) and decreases solute transport and root growth (Assouline, 2006). (Lynch and Bragg 1985). Macropores (pore widths greater than 1,000 m) expand water routes, resulting in increased infiltration and reduced runoff. Surface sealing, on the other hand, minimizes the amount of macropores. Long-term and recurrent cultivation of surface soils during cropping regimes reduces natural aggregation in agricultural soils. Plow pressure, the weight of heavy agricultural machinery, and the hoof strokes of farm or grazing animals can cause soil aggregates to collapse. Finer particles are released as natural aggregates are dissolved, and when exposed, they disperse under raindrop and irrigation action. Particles that have been dispersed reorient and fill in bigger pores. In dry climates and coastal regions, sandy loams, sandy clay loams, and sodic soils are the soils most susceptible to aggregate deterioration. Following cultivation, sodic soils prone to slaking and dispersion are especially vulnerable. Low-organic-matter soils are unable to produce stable peds that can withstand slacking. Heil et al. (1997) found that several Sahelian Alfisols are prone to sealing due to low soil organic matter content and erosion-related exposure of finer-textured subsoil. Their research locations were in the Hamdallaye watershed (500 ha), on six soil series designated as Psammentic Kandiustalf and Petroferric Kanhaplustult, with soil textures ranging from sandy loam to sand. All structural seals studied in the watershed had a 0.1–1.0 mm thick continuous plasmic clay layer within 4 mm of the surface and were morphologically comparable. Organic C concentration at 0–50 mm depth was very low (0.1–0.2%) at sealed sites and slightly higher at open sites. The aggregation was too feeble to resist the assault of raindrops. Clay content, which was likely the main element in seal formation, was related to extractable Fe and Al content of the six-soil series. In most cases, with 90 mm h1 simulated rainfall, the same six soils formed a seal during the initial 30-minute rainfall event, with no change in layer thickness thereafter. Soils containing more than 15% (silt + clay) generated a 2-layer structural seal, whereas soils with a finer texture formed 4-layer structural seals. The kinetic impact of raindrops on the soil surface and the transport of soil particles by flowing water are the physical processes of soil slaking and sealing. The affected soil aggregates break down into primary mineral particles when the drop impact pressures exceed the internal cohesion of the impacted soil aggregates. These particles are washed into the soil surface layer or carried by surface runoff (Schmidt 2010). When translocated particles are deposited, they may clog soil pores, forming surface layers with higher bulk density and lower saturated hydraulic conductivity than the soil underneath (Betzalel et al. 1995). Soil erosion and flood danger are significantly exacerbated as a result of the loss of soil water storage and infiltration capacity.

Surface Crusting

Surface crusts are formed as a result of additional consolidation and drying. Sealing development and crust formation are widespread occurrences in various soils around the world, particularly in arid and semiarid soils. Disintegration, separation, entrainment, deposition, and compression are some of the interactions between water and soils that occur when it rains. These activities lead to the production of seal and, as a result, soil crust. Crust is a thin layer of soil at the surface with higher density, shear strength, and poorer hydraulic conductivity than the underlying soil (Zejun et al. 2002). The mechanism of crust development is comprised of two key processes that work in tandem:

- Disintegration of soil aggregates and compaction of soil particles caused by raindrop impaction are examples of physical activity.
- Physical chemical action, such as aggregate dispersion and soil particle movement, clog conducting pores and generate a less permeable layer in the topsoil region (Cai et al. 1998).

The texture and stability of the soil, the strength and energy of rainfall, gradients and length of slope, and the electrolyte content of the soil solution and precipitation all play a role in the creation of seal and crust (Remley and Bradford 1989).

There are two types of surface crusts:

- structural crust and
- depositional crust.

A structural crust is a thin layer of soil that is more compact than the material beneath it and is a few millimeters to a few centimeters thick. The structural crust does not require the import of external elements to form. Structural crusts can also form as a result of cattle trampling or agricultural machinery traffic. Hard-setting crusts and traffic crusts are examples of structural crusts. The process of compaction and increased bulk density in soil that occurs without the application of an external load is known as hard-setting. Northcote (1960) coined the phrase "hard-setting" in his Australian soil categorization system. Hard-setting is a property of soil horizons with unstable soil aggregates, such as cultivated seedbeds. When the soil is wet, soil aggregates collapse and the seedbed slumps, resulting in a hard, structureless mass of dirt. Hard-setting, on the other hand, necessitates a much greater thickness of material, which frequently includes both the A1 and Ap horizons as well as the E horizon (Lal, 2006). The collapse of some or all the aggregates during and after soaking of previously loosened topsoil is also known as hard-setting. Slumping and uniaxial shrinkage are two physical processes that occur during the hardening process. Slumping happens when a soil containing water-unstable aggregates is wetted and then dried. The aggregates soften and swell at the same time, suspending some or all of the finer particles (silt and clay). The clay parts spread to some extent. Aggregates disintegrate because they lack the strength to sustain the strains induced by rapid water uptake, trapped air, the mechanical action of rapidly moving water (Collis-George and Greene 1979), or differential swelling. Uniaxial shrinkage is essential because it increases the strength of hard-setting soils by bringing particles closer together. Uniaxial shrinkage occurs at least during the early phases of drying, according to laboratory investigations on the behavior of aggregate beds of a hard-setting soil wetted under tension or at zero potential. Starting with a cultivated bed consisting of dry aggregates, Mullins et al. (1987) suggested the following explanation for the formation and increase in strength found in hard-setting soils: When the system is wet, part or all of the silt and clay is mobilized. This can happen as a result of slaking and/or dispersion. The mobilized material is taken behind the retreating water meniscus during the early stages of drying to occupy concavities on the surface of sand grains and any remaining aggregates, establishing annular bridges between them (Kemper et al. 1987). External pressure from agricultural machinery and animals' forms traffic crusts, which can significantly reduce penetrating water and seedling emergence. Surface compaction of damp or moist soils and mechanical disintegration of the surface soil aggregates are two mechanisms that might cause crust development. Some soils are hard-setting both naturally and genetically. In tropical locations, they are plentiful (Fabiola et al. 2003). Hard-setting soils are unable to form water-stable aggregates by nature. In soils with a high percentage of exchangeable salt, a hard setting condition can result. Some hard-set soils are impervious to water, compacted, and cemented. Cementing agents could include amorphous silica and imogolite-like aluminosilicates (Chartres et al. 1989). A depositional crust, on the other hand, forms when soil particles suspended in water settle on the soil surface as water infiltrates or evaporates. Depositional crusts are always built with materials originating from outside sources. Deposition of suspended particles on the soil surface forms depositional crusts. Some cultivated and non-cultivated soils include these. Flood and furrow irrigation water, raindrop impact splash of loose soil particles, overflow and floods from rivers and runoff, and sheet erosion are the main sources of fine eroded soil particles. In turbid suspension, clay and silt particles can either disperse or flocculate. When the electrolyte content in the suspension reaches the clays' flocculation threshold, they flocculate (Oster et al. 1980). The open structure and high permeability of depositional crusts created by flocculated particles. When the suspension electrolyte concentration is below the flocculation threshold, scattered particles settle to form the depositional crust, which has a poorer hydraulic conductivity than the parent soil by many orders of magnitude. Sealing the formation and crust can minimize soil infiltration and enhance runoff at the surface, hence increasing the transport capacity for entraining detached components from soils (Levy et al. 1994). Because the production of sealing and crust is so intimately linked to soil erosion, many scientists have focused on it. Particle detachment and transport are involved in both soil crust and soil erosion processes. Although most erosion models ignore the crust's susceptibility to erosion, new approaches have emerged that involve a clear distinction between detachment and transport mechanisms (Bissonnris et al. 1998). As a result, it is thought that one of the supporting theories for the new soil erosion model should be the basic notion of sealing and crust. Crusting is generally used to destroy soil aggregates. Electrochemical forces bind the clay mineral particles together, which hold aggregates together. Clay mineral kinds, their electrochemical properties, and the electrolyte content of the soil solution all influence aggregate stability. The less vulnerable a soil's aggregate is to crusting, the more stable it is. The following are some of the factors that influence crusting:

***Clay Minerals*:** Because of edge-to-edge and edge-to-face linkages between clusters, kaolinitic soils usually feature stable aggregates, as some positive charges exist on the broken edges, while the large particle faces disclose a negative charge. Kaolinite is a clay that does not expand or swell. As a result, soil aggregates comprised primarily of kaolinite clay do not slake quickly when wet. Montmorillonite, on the other hand, is an expanding clay that expands dramatically when wet. As a result, soils with high montmorillonite content form water-unstable aggregates. When wet, certain soils, such as Vertisols, slake easily. When wet, they become sticky, and when dry, they become extremely rigid.

***Exchangeable Cations*:** Aggregate breakdown and dispersion are involved in the production of all soil crusts. Nature and the distribution of exchangeable cations retained by electrical charges on colloidal surfaces influence the dispersion of soil colloids. Hydrogen (H^+), calcium (Ca^{++}), magnesium (Mg^{++}), and other polyvalent cations are substantially more dispersive than exchangeable sodium (Na^+). To generate flocculation, a solution with a substantially higher concentration of exchangeable Na^+ is required as the % of exchangeable Na^+ increases. In other words, clay with a high exchangeable sodium percentage will disperse rapidly at the soil surface due to the dilution action of rainwater.

***Organic Matter*:** Organic substances may stabilize soil structure in two different ways:

- organic substances can reduce interaction of water with inorganic colloids, and
- it can bind soil particles together physically or chemically.

Humic compounds link particles together like cement. The negative and positive edges of amides, carboxyl, hydroxyl, and other polarizable groups are exposed along the lengthy complicated surfaces of organic polymers.

***Iron and pH*:** Iron oxides have the potential to bind soil particles together. Crusts can form in kaolinitic soils that are high in iron oxides and have a low pH.

Soil Compaction

Soil compaction is the physical consolidation of soil by an applied force, most commonly tillage instruments, which destroys soil structure, compresses soil volume, increases bulk density, reduces porosity, and restricts water and air circulation within the soil. Surface sealing, crusting, and hard-setting are all examples of soil compaction caused by slaking of soil structure in the surface soil. Soil consolidation and compaction are sometimes distinguished. Soil consolidation is defined as a process in which a saturated soil is compacted over time under a long-term load, resulting in a decrease in porosity and water ejection. Soil compaction, on the other hand, is defined as a process in which an unsaturated soil is crushed by a load applied for a brief period with no water evaporation. However, in the case of compacted soil management, such a differentiation does not appear to be feasible. Soil compaction manifests itself as stunted plant growth, shallow and deformed plant roots, standing water, the creation of enormous clods following tillage, and physically dense soil. The EU Soil Framework Directive lists soil compaction as one of the five hazards to long-term soil quality (Commission of the European Communities 2006). Globally, there are around 68.3 million hectares of compacted soil, accounting for 4% of human soil deterioration (Oldeman et al. 1991). Compaction accounts for around 17% of the overall degraded area in Europe. It's a complicated problem with a lot of machine/soil/crop/weather linkages that could have economic and environmental implications for global agriculture (Soane and van Ouwerkerk 1995). Soil compaction is a global issue that is often linked to mechanized agriculture. It has resulted in yield decreases of 25–50 percent in several European and North American regions (Eriksson et al.

1974) and between 40 and 90 percent in West African countries (Kayombo and Lal 1994). In the United States, on-farm losses due to soil compaction are estimated to be about $1.2 billion each year (Gill 1971). Compaction, according to Oldeman et al. (1991), is by far the most common type of physical deterioration of agricultural soils.

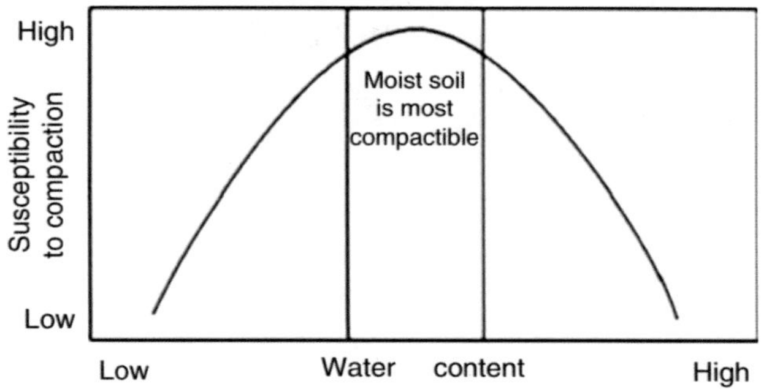

Figure 8: Soil moisture level and susceptibility to compaction

Source: Osman, (2014).

Soil compaction can be classified into two types: shallow and deep. Normal tillage can break up shallow compaction that occurs near the soil surface. Shallow soil compaction includes surface crust and hard-setting. Axle load is the primary source of deep compaction. Tillage pans or plow pans are compacted layers below the plow depth. According to Van den Akker and Schjonning (2004), significant regions have been subjected to subsurface compaction as a result of the usage of heavy farm machinery. Deep compaction can go down to 75 centimeters below the surface. Because it is below the regular tillage zone, it is extremely difficult to fix. Deep compaction lowers water and air storage in the deeper part of the soil profile, stifling crop growth of all kinds. Deep compaction makes it difficult for deep-rooted crops like corn and alfalfa to develop a complete root system. Soil compaction is caused by any process that reduces pore space. When driving over a field, farm machinery such as tractors and other equipment can compact soil. Soil can be compacted by livestock and even humans. The amount of soil compaction is determined by several factors, including soil physical conditions, load weight and design, load distribution across the number of axles and tires, and, to a lesser extent, the number of trips made over the field. The texture, structure, organic matter, and water content of the soil all have a role in determining its susceptibility to compaction. Soils with similar-sized particles compact less than soil with a variety of particle sizes. Smaller particles can fill the holes between larger particles, causing the soil to become denser. A sandy loam soil can be compacted more easily than a sandy or clayey soil. The structure of a soil with a higher degree of organic matter is generally better. Compaction affects hard, dense, low-organic-matter soils more than loose, friable, high-organic-matter soils. Dry earth is difficult to compact. Soil compaction rises when the water content of the soil rises during tilling. Because water occupies the pores and is not compressible, a saturated soil resists compaction. When soil water content is somewhat higher than field capacity, soils are easily compacted. Traffic over the fields, especially with large farm machinery, is a primary cause of soil compaction. Tractors, combine harvesters, and another farm machinery are enormous and heavy. Tractors can weigh up to 20 tons when fully laden, and liquid manure spreaders can weigh up to 15 tons. To sustain such loads, higher tire inflation pressures are used. The soil surface is subjected to increased pressures as a result of this heavyweight. With radial tires, soil pressure is roughly equal to inflation pressure, however with bias-ply tires, soil pressure is higher towards the tire's edge because the sidewall carries part of the load. Tractors, combines, and implements employ both radial and bias-ply tires. When compared to single tractor tires, dual tractor tires reduce soil pressure while increasing the affected area. Soil compaction can be caused by as little as 4 psi pressure; duals or even high-flotation tires do not prevent compaction; they simply shift the distribution. A pickup truck exerts roughly 50 psi on the earth, a liquid manure spreader 70–90 psi, and a 500 kg cow about 40 psi. In terms of soil compaction, the number of passes made by machinery across the field is critical. In bad weather, even a single pass over a field might inflict severe damage. Eighty percent of the possible compaction is caused by the initial wheel pass. Subsequent passes compact the soil even more, albeit to a lesser extent. The additional compaction becomes very modest after four passes. Soil compaction can be caused by tillage or alleviated by it. Soil is broken up into smaller particles during tillage activities. The structure that provides desirable pore space is destroyed by excessive tillage. Some tillage equipment, such as moldboard plows, can aerate the soil and improve percolation at the surface while also forming a compacted layer just below the plowing depth. A "plow-pan" or "hard-pan" is a type of layer like this. Disks can also over-till the soil near the surface, resulting in a hard pan just below tillage depth, especially if multiple passes are done.

Soane et al. (1982), Lipiec and Stepniewski (1995), and Alakukku (1999) evaluated the impact of soil compaction on soil characteristics and processes (1999). Soil compaction has been discovered to have varying degrees of impact on practically all physical, chemical, and biological properties and processes of soil. Soil compaction alters the pore size distribution of mineral soils by lowering porosity, particularly macro-porosity (diameter >30 m; Eriksson 1982; Ehlers 1982). Compaction changes the pore geometry, continuity, and morphology in addition to the volume and quantity of macropores. Many soil qualities linked to soil working, drainage, agricultural growth, and the environment are negatively impacted by soil compaction. Field traffic compacts soils, reducing bulk density (Arvidsson 1998), shear strength, and penetrometer resistance (Blackwell et al. 1986), restricting root growth and increasing tillage draft requirements. Soil compaction lowers infiltration and saturation hydraulic conductivity (Pietola et al. 2005). (Alakukku et al. 2003). Compaction of the soil inhibits CO2 and O2 exchange (Simojoki et al. 1991). Drainage issues arise as a result of subsoil compaction, which reduces permeability. Waterlogging can be caused by soil compaction. Poorly drained soil might also take a long time to dry, limiting the number of days available for field work. By increasing denitrification, the reduction in drainage rate related to soil compaction might be expected to enhance greenhouse gas emissions from soil (Ball et al. 1999). Surface runoff and topsoil erosion are exacerbated by compaction (Fullen 1985). Soil compaction affects crop development, yield, and water and fertilizer use efficiency through changing soil properties and processes. Compaction of the soil affects yield (Hanssen 1996), crop water usage efficiency (Radford et al. 2001), and nutrient uptake (Radford et al. 2001). (Alakukku 2000). Low seed germination, reduced seedling emergence, lower root number and length, restriction of downward penetration of the main root axes, decrease in leaf thickness, rise in shoot-to-root ratio, and decrease in crop grain production are all typical plant responses to soil compaction (Fageria et al. 2006). The degree to which root growth is restricted in tight soil is also determined by the plant's species and age (Masle 2002). Reduced rooting volume is the primary cause of stunted plant growth (Grzesiak et al. 2002; Masle 2002). Because high soil impedance effects root elongation and proliferation, the restricting effect of soil compaction can be physical and physiological restrictions to total plant growth and yield through inadequate root system development (Iijima et al. 1991; Grzesiak et al. 2002). The size of the root zone and the spread of roots are the most important factors in water and mineral nutrient acquisition.

Waterlogging

Waterlogging is a condition in which the soil is saturated with water for an extended period. Soils can be saturated with water for a brief period, say a few hours, as in the case of flash floods in piedmont locations. This isn't considered waterlogging because it's only transitory. This water is simply drained. Waterlogging is a soil situation in which draining excess water is difficult and time, labor, planning, and energy are all required. Waterlogging can be caused by a variety of factors, including natural events and human actions. Heavy rains, low lands, clay soils, flooding, and the existence of impervious subsoil are all natural phenomena. Faulty irrigation, insufficient drainage, surface sealing, and deep soil compaction by tillage instruments are all examples of human actions. Wetlands are constantly flooded. Waterlogging can also happen when the groundwater table rises to the root zone and stays there for a long time (Michael and Ojha 2006). Waterlogging is a type of standing water in the farm that does not go away with time (Murty 1985). Waterlogging affects around 10% of all irrigated land on the planet. As a result, productivity in these croplands has dropped by around 20%. Heavy rains, low lands, clay soils, flooding, and the existence of impervious subsoil are all natural causes of waterlogging. Faulty irrigation, insufficient drainage, and surface sealing, as well as deep soil compaction by tillage instruments, are all human actions that cause or aggravate waterlogging. Wetlands are permanently waterlogged, but we're mostly concerned about arable lands since waterlogging slows the growth of most crop species. The most common causes of waterlogging are compaction-induced poor drainage and an increase in the groundwater table. Organic matter depletion, structure deterioration, dispersion, compression, and consolidation all contribute to compaction, which includes surface sealing, crusting, hardening, and deep compaction. These processes arise as a result of improper tillage and the pressure exerted by farm machinery. Compaction of the soil affects hydraulic conductivity, which reduces infiltration and percolation. Rainwater or irrigation water cannot reach the regolith's deep groundwater table and drain away. In the absence of considerable base flow, water collects and the groundwater table rises toward the surface of the soil. In tiny agricultural holdings, a limited area is usually irrigated, resulting in over-irrigation and significant percolation, which causes the groundwater table to rise. Root respiration and function are hampered by rising groundwater levels. Salinity can be created by an elevated groundwater table, which reduces crop output. Plant roots are limited to the loose surface soil in soils with heavy compaction, and extra irrigation water cannot easily penetrate through the impermeable layer. It results in a shallow root system and waterlogging. An impermeable stratum can form beneath the top layers of pervious soils in some instances. Water seeping through pervious soils will not be able to travel deep in this instance, and as a result, waterlogging will occur soon. In the lack of adequate natural drainage, well-designed artificial drainage systems are required. Drainage systems frequently fail to perform satisfactorily because they are neither sufficiently deep nor sufficiently frequent. Aeration is harmed by waterlogging. The oxygen concentration in soil may be fully depleted within a few days of floods. Changed physiological processes caused by low oxygen and reduced root respiration in wet soils impair crop plant growth and yield. Crop failure is frequently caused by waterlogging. Furthermore, waterlogging makes it impossible to stick to a strict crop calendar. Normal agricultural procedures like tilling and plowing are difficult to carry out on damp soils. Waterlogging causes delays in plowing, sowing, and planting. Many croplands become wastelands as a result of waterlogging. Dissolved salts in groundwater are carried to the surface soil by rising water tables. Water evaporates continuously, leaving salts in the soil. As a result of the high-water table, soil salinity develops, which is difficult to reclaim. Salinity in the soil diminishes productivity. Due to poor irrigation in various nations, including Pakistan, many soils have turned salty and unfit for cultivation.

Lowering of Groundwater Table

Groundwater is water that is found beneath the earth's surface in pores and fractures in soil and rock. The groundwater table is the highest level of this subsurface water. The water table is the point where the unsaturated and saturated zones meet. It symbolizes the groundwater's upper surface. The water level in abandoned wells is frequently the same as the water table. Up to the groundwater table, the pore spaces in soil become totally saturated with water. Groundwater recharge is the process of water entering the saturated zone beneath the ground. Water travels downward from surface water to groundwater in a hydrologic process known as groundwater recharge. This process normally takes place below plant roots in the vadose zone, and it's often expressed as a flux at the water table's surface. Natural and manmade factors both contribute to recharge. Natural and human-caused phenomena are both sources of recharge for a groundwater system. Recharge from precipitation, wetlands, lakes, ponds, and rivers (including permanent, seasonal, and ephemeral flows), as well as from other aquifers, are all natural sources. Irrigation losses from canals and fields, leaking water mains, sewage, septic tanks, and over-irrigation of parks, gardens, and other public amenities are all human-induced sources of recharge. Direct recharge from precipitation percolation and indirect recharge from runoff ponding are two types of recharges from these sources. The groundwater table rises when groundwater recharge exceeds discharge. The groundwater table is lowered when groundwater pumping and discharge exceeds recharging. Groundwater discharge is the process of removing water from the groundwater. Some, and frequently a large portion, of the water in rivers comes from groundwater seepage into the streambed. In most physiographic and climatic contexts, groundwater contributes to streams. The amount of stream water that originates from groundwater input varies depending on the topography, geology, and climate of a given place. Groundwater discharge is the movement of groundwater in certain directions. It is the process of extracting water from the groundwater. In urban and agricultural areas, a lot of groundwater is withdrawn from shallow and deep tube wells. Excessive groundwater pumping is the most serious consequence of lowering the water table. This puts a future water scarcity for urban and agricultural use in jeopardy. Groundwater pumping has the effect of reducing groundwater levels below the depth required for streamside or wetland plants to exist. Overall, riparian vegetation and wildlife habitat are being lost. Land subsidence is caused by a lack of support below ground. The soil collapses, compacts, and descends as water is removed from it. This is determined by several factors, including the type of soil and rock beneath the surface. As a result, the water table beneath the soil surface should be kept at a desirable depth. By limiting oxygen, a shallow water table limits root growth and elongation. The lowering of the water table on a large scale has been identified as a major contributor to desertification.

Subsidence of Organic Soils

Subsidence is the drop in ground level caused by organic matter compression, consolidation, compaction, and oxidation, especially as a result of draining and cultivating peat soils (Histosols). In many regions of the world, subsidence has long been a major issue in the reclamation and development of peat fields. Aeration is improved by draining the surface soil layers above the submergence level. As a result, peat soil reclamation will always be accompanied by oxidation of these layers. Reclamation of peat fields in the Middle Ages in the Netherlands has steadily subsided from 0.5 m above mean sea level to 1–2 m below. Despite a steady modest drainage, the land subsided by around 2 meters within a duration of 8–10 centuries (Nieuwenhuis and Schokking 1997). The oxidation of organic waste is responsible for 85 percent of this subsidence, which will continue at a rate of 5 cm each year1 (Schothorst 1982). The most common response of Histosols to drainage for agricultural purposes is land subsidence. Subsidence rates in drained peaty areas around the world range from less than 1 centimeter per year1 to more than 10 cm per year1. Between 1924 and 1978, the arable organic soils in Florida suffered an average subsidence rate of 2.5 cm per year1 (Ingebritsen et al. 1999). Between 1922 and 1950, histosols in California settled at a pace of up to 8 cm per year1 (Rojstaczer and Deverel 1995). Land subsidence rates in Malaysia declined from 12 cm per year1 from 1960 to 1974 to 6.4 cm per year1 for the next 14 years, then to 2 cm per year1 subsequently, according to records (Wosten et al. 1997). Following drainage, Stephens et al. (1984) identified at least five mechanisms of organic soil subsidence: desiccation-induced shrinkage, consolidation, wind and water erosion, fire, and biochemical oxidation. In temperate and tropical peat soils, the latter has been discovered to be the primary source of land sinking (Deverel and Rojstaczer 1996). Buildings and structures in low-bearing-capacity locations must be erected on piling foundations. The relative height of piled buildings and structures in relation to the surrounding environment will alter due to subsidence. The regions around these structures and buildings will need to be raised from time to time with earth or other fill material. When it comes to connecting utilities, certain precautions must be taken (power lines, water mains, etc.). The design height of embankments must be modified on soils with soft clay or peat layers to account for potential subsidence; otherwise, the safety criteria may not be met. Clay composition, depth of peat layer, loss of water owing to drainage, water table, and load of machineries, according to Segeren and Smits (1980), all affect the rate and amount of subsidence. Compared to sandy soils, clayey sediments lose more water. As a result, clay soils will sink more. At a given clay content, the sinking produced by the shrinkage of distinct soil layers diminishes with depth. The subsidence will be larger as the thickness of the compressible layers increases. The peat soils are significantly compacted when they are cultivated with heavy machinery.

Desertification

Desertification is a land degradation process that happens primarily in arid and semiarid regions as a result of human activity on ecosystems paired with harsh climate conditions. "Desertification is the depletion of terrestrial ecosystems under the action of man," defined Dregne (1977). Reduced production of desirable plants, unwanted changes in biomass and diversity of micro and macro flora and fauna, rapid soil deterioration, and increasing risks for human occupation are all signs of deterioration in these ecosystems." Desertification is defined as "land deterioration in arid, semi-arid, and dry sub-humid areas as a result of many factors, including climate fluctuations and human activities," according to the United Nations Convention on Desertification (UNCCD), which was enacted in 1994. Desertification transforms woods, grasslands, shrublands, and croplands into desert-like landscapes. Since the 1970s, the international community has become more concerned about desertification's destructive and destabilizing effects on the natural environment and human culture. According to UN estimates, the occurrence has piqued the curiosity of 70 percent of neighboring dry lands, or roughly 30 percent of the world's cultivable territory. The problem is notably acute in Africa, as well as various developing countries in Asia, South America, and the Caribbean; nevertheless, it also affects the United States, Australia, and Southern Europe (particularly Greece, Spain, Portugal, and Italy) (Perini et al. 2009). According to the OECD, erosion, submersion, acidification, salinization, soil compaction, surface crusts and compact layers along the profile, loss of organic matter, deterioration of the soil structure, accumulation of hazardous compounds, and loss of nutrients are the main causes of soil degradation (Perini et al. 2009). Desertification has an impact on over 2 billion people in over 100 nations on all five continents (Bied Charreton 2008). Desertification destroys ten million hectares of productive land each year, resulting in the loss of 24 billion tons of topsoil. Desertification has three major consequences: a loss of biological productivity, a loss of economic productivity, and a loss of landscape complexity (UNCCD 1994).

Desertification is caused by a variety of factors that are complicated, often localized, and differ from one region to the next. The severity of the impact varies as well, with less-developed countries suffering more human anguish. Desertification's origins and processes are hotly discussed by experts from a variety of fields, including geography, ecology, environmental science, economics, and sociology. Desertification, according to some experts, is a natural phenomenon that occurs as a result of irregular short- and long-term drought, as seen in the African Sahel (Hill and Peter 1996). Others argue that desertification is a human-caused phenomena caused by land resource pressures, poor development strategies, and land exploitation (Graetz 1996). Desertification is most likely to occur in undeveloped, impoverished areas with limited biomass resources. In dry and semiarid areas, soil and vegetation are naturally vulnerable. Human strain on vegetation and land continues to be immense, but technology, soil, and climate support is insufficient. According to Culet (2002), overgrazing, deforestation, and other agricultural activities may contribute 35, 30, and 28 percent to desertification, respectively. Desertification induced by human influences has a long history; evidence of such degradation can be seen in ancient Mesopotamia near the Tigris and Euphrates rivers. Desertification is influenced by human activities such as deforestation, shifting cultivation, overgrazing, groundwater depletion, frequent burning, and soil compaction. These factors have an impact on desertification, which is a continual and progressive degradation of ecosystems in which plants and animals, as well as other natural resources like water and soil, are strained beyond their ability to adapt to changing conditions. Desertification is defined by the progressive loss of mature, ecosystem-stabilizing flora, as well as the loss of agricultural crop cover and the subsequent loss of unconsolidated soils. In semiarid environments, early human pastoralists moved their small herds of domestic animals in response to food and water availability. Overgrazing of the delicate plant cover was prevented by such regular stock movement. The risk of cattle overgrazing rose as human settlements became more established.

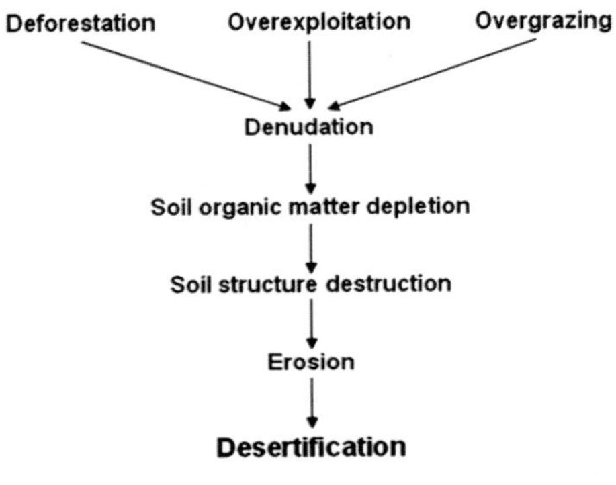

Figure 9: Desertification Scheme

Source: Osman, (2014).

Desertification is mostly caused by overgrazing in many parts of the world. In many locations, the cultivation of marginal lands, which have a high chance of crop failure and a low economic return, increases the risk of desertification. Denudation is caused by the destruction of vegetation in arid regions for fuelwood and the burning of vegetation in numerous places. In dry environments, improper irrigation procedures can lead to salinization, which can inhibit plant growth. Soil organic matter is depleted as a result of decreased biomass production and increased breakdown. Water and wind erosion erode the soil structure and remove soil particles. Continuous erosion reveals the subsoil, which is unable to support adequate vegetation. Gullies and sand dunes form across the deflated ground surface as a result of wind and water erosion.

Urbanization

Globally, urbanization is accelerating, and land surfaces are gradually being sealed across the world. Soil serves a variety of roles, and its ecological service, particularly its ability to buffer and purify pollutants, is critical in metropolitan areas. However, soils in metropolitan areas are frequently prone to fundamental changes, if not complete degradation, as a result of different intensive human activities. Their degeneration is essentially a sacrifice process in which they provide ecological functions, such as serving as a pollutant sink, at the expense of their own quality. As a result, they frequently encounter several environmental issues. Physical degradation, such as compaction and structural disintegration, can impair urban soils' ability to infiltrate water and store it, resulting in greater runoff and pollution loads in receiving water bodies. Compacted soil is also thought to exacerbate the metropolitan heat island effect by reflecting more radiation. The major concerns of the urban soil environment are the enrichment of numerous waste items linked with human activities, such as nutrient elements, heavy metals, and organic pollutants. The key characteristics of urban soil contamination include substantial accumulation of so-called "urban elements" like Cu, Zn, Pb, and Hg, but less accumulation of other heavy metals, and spatial isolation, which implies the contamination is not spatially continuous. Heavy metal pollution of urban soils occurs during the development process not only today, but also in the past, especially when rudimentary mining and metal processing were prevalent. Furthermore, organic pollutants, particularly polycyclic aromatic hydrocarbons (PAHs), are frequently found in urban soils, with roadsides and industrial sites being the most vulnerable regions. The presence of 2-6 rings in PAHs, with a predominance of 2-4 rings, strongly suggests a pyrogenic background, such as traffic emissions, industrial operations, and coal combustion.

LAND DEGRADATION PROCESSES AND CLIMATE CHANGE

While the tight interconnection of different systems makes subdivision difficult, it does provide a helpful environment for identifying the most critical 'focal areas' of climate change pressures on land degradation. All types of erosion and SOM declines (soil focus), salinization, sodification, and permafrost thawing (soil/water focus), waterlogging of dry ecosystems and drying of wet ecosystems (water focus), and a broad group of biologically-mediated processes like woody encroachment, biological invasions, and peat bog thawing (water focus) are among the land degradation processes responding more directly to climate change pressures. Ground subsidence can be influenced indirectly by climate change due to rising sea levels (Keogh and Törnqvist 2019). Even when climate change has a direct impact on degradation processes, it can be a secondary driver overshadowed by other major human influences. Three processes in which climate change is a dominant global or regional pressure and the main driver of their current acceleration are notable exceptions. Coastal erosion as a result of rising sea levels and increased storm frequency/intensity (high agreement, moderate evidence) (Johnson et al. 2015). increased burning in response to warming and changing precipitation regimes (high agreement, substantial evidence) (Batir et al. 2017); and permafrost thawing in response to warming (high agreement, robust evidence) (Batir et al. 2017). (Taufik et al. 2017) Climate change not only exacerbates many of the well-known ongoing land degradation processes in managed ecosystems (i.e., croplands and pastures), but it also becomes a dominant pressure that introduces novel degradation pathways in natural and seminatural ecosystems, according to the previous assessment. Climate change has altered species invasions and the destruction they produce by increasing invasive species' movement, colonization, establishment, and ecological effect, as well as weakening their control strategies (medium agreement, medium evidence) (Hellmann et al. 2008).

Table 16: Major land degradation processes and their connections with climate change.

Process	**Climate change Pressure**	**Specific Impact**
Wind erosion	Altered wind/drought patterns. Indirect effect through vegetation type and biomass production shifts.	Radiative cooling by dust release. Ocean and land fertilization and carbon burial. Albedo increase. Dust effect as condensation nuclei.
Water erosion	Increasing rainfall intensity Indirect effects on fire frequency/ intensity, permafrost thawing, biomass production.	Net carbon release. Net release is probably less than site-specific loss due to deposition and burial. Albedo increase.
Coastal Erosion	Sea level rise, increasing intensity/ frequency of storm surges.	Release of old buried carbon pools.
Subsidence	Indirect through increasing drought leading to higher ground water use. Indirect through enhanced decomposition (e.g., through drainage) in organic soils.	Unimportant in the case of groundwater depletion. Very high net carbon release in the case of drained peatlands
Compaction/ hardening	Indirect through reduced organic matter content.	Contradictory effects of reduced aeration on N_20 emissions
Nutrient depletion	Indirect (e.g., shifts in cropland distribution, BECCS)	Net carbon release due shrinking SOC poofs. larger reliance on soil liming with associated C_2O releases
Acidification/ overfertilization	Indirect (e.g., shifts in cropland distribution, BECCS). Sulfidic wetland drying due to increased drought as special direct effect.	N_2O release from overfertilized soils, increased by acidification. Inorganic carbon release from acidifying soils.
Pollution	Indirect (e.g., increased pest and weed incidence)	Unknown, probably unimportant

Organic matter decline	Warming accelerates soil respiration rates. Indirect effects through changing quality of plant litter or fire/waterlogging regimes	Net carbon release.
Salinization	Sea level rise. Water balance shifts. Indirect effects through irrigation expansion	Reduced methane emissions with high sulfate load. Albedo increase.
Metal toxicity	Indirect	Unknown, probably unimportant
Sodification (increased sodium and associated physical degradation in soils)	Water balance shifts. Indirect effects through irrigation expansion	Net carbon release due to soil structure and organic matter dispersion. Albedo increase.
Permafrost thawing	Warming, seasonality shifts and accelerated snow melt leading to higher erosivity.	Net carbon release. CH_4 release
Waterlogging of dry systems	Water balance shifts. Indirect effects through vegetation shifts	CH_4 release. Albedo decrease
Drying of continental waters/ wetland/ lowlands	Increasing extent and duration of drought. Indirect effects through vegetation shifts	Net carbon release. N_2O release. Albedo increase.
Flooding	Sea level rise, increasing intensity/ frequency of storm surges, increasing rainfall intensity causing flash floods.	CH_4 and N_2O release. Albedo decrease
Eutrophication of continental waters	Indirect through warming effects on nitrogen losses from the land or climate change effects on erosion rates. Interactive effects of warming and nutrient loads on algal blooms.	CH4 and N2O release.
Woody encroachment	Rainfall shifts (medium confidence on effects and trends), CO rise.	Net carbon storage. Albedo decrease
Species loss, compositional shifts	Habitat loss as a result of climate shifts	Unknown

Soil microbial and meso-faunal shifts	Habitat loss as a result of climate effects shifts.	Unknown
Biological soil crust destruction	Warming. Changing rainfall regimes. Indirect through fire regime shifts and/or invasions.	albedo Radiative cooling through rise and dust release.
Invasions	Habitat gain as a result of climate Shifts.	Unknown
Pest outbreaks	Habitat gain and accelerated reproduction as a result of climate shifts.	Net carbon release

Source: Olsson et al., (2019).

Chapter 3

Soil Health Management

CONTENTS

INTRODUCTION

Soil health refers to a soil's ability to function as a living ecosystem that supports plants, animals, and humans. Because only living things can have "health," seeing soil as a living, breathing ecosystem represents a transformation in how we think about and manage our country's soils. Soil isn't an inert growing medium; it's home to billions of bacteria, fungus, and other species who work together to form a complex symbiotic ecosystem. By cycling nutrients, absorbing, draining, and retaining rainwater and snowmelt for use during dry periods, filtering, and buffering water to remove potential pollutants, and providing habitat for the soil biological population to flourish and diversify, this ecosystem can be managed to support plants and animals. Soil is home to one of the most diverse assemblages of living species on the planet, all of which are intricately linked by a complex food web. Depending on how it is maintained, it might be either sick or healthy. The great diversity of its biota and the large concentration of non-living soil organic matter are two important properties of a healthy soil. It is reasonable to infer that a soil is healthy if the organic matter is added or maintained at an acceptable level for profitable crop development. Infestations of soil-borne pests are less likely in healthy soil. Striga, for example, is a parasitic weed that thrives in good soils. In fertile soils, even pests that aren't found in the soil, such as maize stem borers, inflict less damage. In the tropics, soil biota diversity is larger than in temperate zones. Agri-ecosystems in the tropics are particularly vulnerable to soil degradation since the rate of agricultural intensification will be higher in the future. Subsistence farmers in the tropics are more reliant on these processes and services than farmers in other regions, thus any loss of biodiversity and, ultimately, ecosystem functioning will harm them more than farmers in other regions. The ability of a soil to store and release nutrients and water to plants, as well as encourage and sustain plant growth, is determined by the functional interactions of soil biota with organic and inorganic components, air, and water. Large stores of stored nutrients do not guarantee excellent soil fertility or crop yield. Because plants absorb most of their nutrients in a water-soluble form, nutrient transformation, and cycling – through biological, chemical, and physical processes – is critical. The nutrients must be carried to the plant roots by free-flowing water. Soil structure, which impacts a soil's water-holding capacity and rooting depth, is another important component of a healthy soil. Physical limits, such as a high-water table, bedrock, or other impermeable strata, as well as chemical issues such soil acidity, salinity, sodality, or poisonous chemicals, might limit rooting depth. Crop productivity might be hampered by a lack of any of the 15 nutrients essential for plant growth. It is critical to ensure their availability in soils and to apply a balanced amount of nutrients from organic sources and mineral fertilizers, if necessary, to achieve the increased productivity required to fulfill present and future food demand. Where deficits arise, timely supplementation of micronutrients in "enriched" fertilizers could be a source of improved crop nutrition. N-fixing legumes and trees can also be used into farming systems to supply nitrogen to the soil. Trees and some soil-improving legumes can pump nutrients up from the subsoil that would otherwise never reach crops because of their deep roots. Other biological relationships can improve crop nutrition, such as those between crop roots and soil mycorrhizae, which help cassava acquire phosphate in deficient soils. When ecosystem processes fail to provide enough nutrients for high yields, intensive farming will rely on the prudent and efficient use of mineral fertilizers. A sustainable soil health management system is built on a combination of ecosystem processes and prudent mineral fertilizer use. This system can produce higher yields while utilizing fewer external inputs.

THE SOIL FOOD WEB

Soil health refers to a soil's ability to function as a living ecosystem that supports plants, animals, and humans. Because only living things can have "health," seeing soil as a living, breathing ecosystem represents a transformation in how we think about and manage our country's soils. Soil isn't an inert growing medium; it's home to billions of bacteria, fungus, and other species who work together to form a complex symbiotic ecosystem. By cycling nutrients, absorbing, draining, and retaining rainwater and snowmelt for use during dry periods, filtering, and buffering water to remove potential pollutants, and providing habitat for the soil biological population to flourish and diversify, this ecosystem can be managed to support plants and animals. Soil is home to one of the most diverse assemblages of living species on the planet, all of which are intricately linked by a complex food web. Depending on how it is maintained, it might be either sick or healthy. The great diversity of its biota and the large concentration of non-living soil organic matter are two important properties of a healthy soil. It is reasonable to infer that a soil is healthy if the organic matter is added or maintained at an acceptable level for profitable crop development. Infestations of soil-borne pests are less likely in healthy soil. Striga, for example, is a parasitic weed that thrives in good soils. In fertile soils, even pests that aren't found in the soil, such as maize stem borers, inflict less damage. In the tropics, soil biota diversity is larger than in temperate zones. Agri-ecosystems in the tropics are particularly vulnerable to soil degradation since the rate of agricultural intensification will be higher in the future. Subsistence farmers in the tropics are more reliant on these processes and services than farmers in other regions, thus any loss of biodiversity and, ultimately, ecosystem functioning will harm them more than farmers in other regions.

The ability of a soil to store and release nutrients and water to plants, as well as encourage and sustain plant growth, is determined by the functional interactions of soil biota with organic and inorganic components, air, and water. Large stores of stored nutrients do not guarantee excellent soil fertility or crop yield. Because plants absorb most of their nutrients in a water-soluble form, nutrient transformation, and cycling – through biological, chemical, and physical processes – is critical. The nutrients must be carried to the plant roots by free-flowing water. Soil structure, which impacts a soil's water-holding capacity and rooting depth, is another important component of a healthy soil. Physical limits, such as a high-water table, bedrock, or other impermeable strata, as well as chemical issues such soil acidity, salinity, sodality, or poisonous chemicals, might limit rooting depth. Crop productivity might be hampered by a lack of any of the 15 nutrients essential for plant growth. It is critical to ensure their availability in soils and to apply a balanced amount of nutrients from organic sources and mineral fertilizers, if necessary, to achieve the increased productivity required to fulfill present and future food demand. Where deficits arise, timely supplementation of micronutrients in "enriched" fertilizers could be a source of improved crop nutrition. N-fixing legumes and trees can also be used into farming systems to supply nitrogen to the soil. Trees and some soil-improving legumes can pump nutrients up from the subsoil that would otherwise never reach crops because of their deep roots. Other biological relationships can improve crop nutrition, such as those between crop roots and soil mycorrhizae, which help cassava acquire phosphate in deficient soils. When ecosystem processes fail to provide enough nutrients for high yields, intensive farming will rely on the prudent and efficient use of mineral fertilizers. A sustainable soil health management system is built on a combination of ecosystem processes and prudent mineral fertilizer use. This system can produce higher yields while utilizing fewer external inputs.

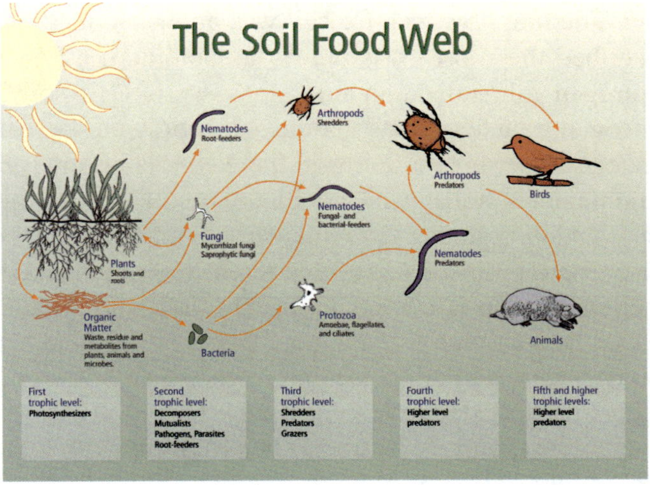

Figure 10: The soil food web

Source: https://www.nrcs.usda.gov/Internet/FSE_MEDIA/nrcs142p2_049822.jpg

Types of Food Webs

Topological (or conventional) food webs, flow webs, and interaction webs are the three forms of food web representations. These webs can be used to describe both above and below ground systems.

Topological webs: Early food webs were topological, descriptive, and presented a nonquantitative depiction of consumers, resources, and their connections. These webs are defined by Pimm et al. (1991) as a map of which creatures in a society devour which others. In 1912, the first topological food web was created to study the predators and parasites of the cotton boll weevil (reviewed by Pimm et al. 1991). Researchers measured the web's interaction chain lengths and connectedness to assess and compare topological webs between ecosystems. One issue with standardizing such measures is that there are frequently too many species to assign each to its own box. The number of species aggregated or split into functional groupings varies depending on the source. Some creatures may even be removed by the authors. The dead material migrating back to detritus is not depicted by convention since it would complicate the graphic, but it is factored into any calculations.

Flow webs: Miosis builds on interconnected food chains by incorporating quantitative data on carbon or other nutrient flow from farmers to consumers. The first flow web for soil was published in 1987 by Hunt et al., who described the short grass prairie in Colorado, USA. The authors evaluated nitrogen mineralization rates for a variety of soil species and predicted nitrogen transferal rates via the soil food web. Researchers from the Lovinkhoeve Experimental Farm in the Netherlands analyzed the flow of carbon and depicted transfer rates with arrows of various thicknesses in another groundbreaking study. A topological web must be built before a flow web may be created. Following the selection of web members, the biomass of each functional group is computed, usually in kilograms of carbon per hectare. Researchers assume that the population of the functional group is in equilibrium when calculating feeding rates. At equilibrium, the rate at which members are lost due to natural death and predation is balanced by the rate at which the group reproduces. The efficiency with which nutrients are transformed into organism biomass can be estimated after the feeding rate is known. This stored energy represents the amount of energy that can be passed on to the next trophic level. Researchers determined that nutrients and energy traveled from lower resources to higher trophic levels through three main channels after constructing the first soil flow webs. The bacterial and fungal channels had the most energy flow, while the herbivory channel, in which creatures ate plant roots directly, had the least. Bacteria and fungi are now well recognized as crucial players in the breakdown of carbon and nitrogen, and they play important roles in both the carbon and nitrogen cycles.

Interaction web: The arrows in an interaction web, seen above right, depict how one group influences another rather than the passage of energy or materials, as in a topological web. Every link in an interactive food web model has two direct effects: one on the resource and one on the consumer. The resource has a beneficial influence on the consumer (the consumer gets to eat), while the consumer has a negative effect on the resource (it is eaten). These trophic effects can have indirect consequences. Dashed lines depict indirect effects, which are the impacts of one element on another that is not directly linked to it. When the predator consumes the root herbivore, the plant consumed by the herbivore may increase in biomass, as seen in the simple interaction web below. As a result, we can conclude that the predator has a positive indirect effect on the plant roots.

Food Web Control

***Bottom-up effects*:** When the density of a resource influences the density of its consumer, this is known as a bottom-up effect. In the diagram above, for example, an increase in root density leads to an increase in herbivore density, which leads to an increase in predator density. Bottom-up control is supported by correlations in abundance or biomass between users and their resources. The relationship between herbivores and plant primary productivity is a well-known example of a bottom-up impact. Herbivores and detritivores' biomass grows in terrestrial environments as primary productivity rises. Increased primary productivity will result in more leaf litter entering the soil ecosystem, providing more nutrients for bacterial and fungal communities to develop. An increase in bacterial and fungal feeding nematodes, which are devoured by mites and other predatory nematodes, will result from more microorganisms. As more resources are added to the base, the entire food web expands. Bottom-up control is a term used by ecologists to describe how higher trophic levels rely on resources from lower trophic levels for biomass, abundance, and diversity.

Top-down effects: Top-down control ideas are significantly more difficult to assess. When a consumer's population density impacts the population density of its resource, such as when a predator affects the density of its prey, top-down effects occur. As a result, top-down control refers to situations in which the abundance, variety, or biomass of lower trophic levels is influenced by consumer effects at higher trophic levels. The indirect impacts of predators are described by a trophic cascade, which is a sort of top-down interaction. Predators cause impacts that cascade down the food chain, affecting the biomass of species at least two links distant in a trophic cascade. In ecology, the role of trophic cascades and top-down management in terrestrial ecosystems is hotly discussed, and the question of whether trophic cascades exist in soils is no less complicated. Trophic cascades can be seen in both bacterial and fungal energy pathways. However, because many other research demonstrates no top-down impacts of predators, cascades may be uncommon. Microbes consumed by nematodes grew quicker when they were grazed on often in Mikola and Setälä's (1998) study. When the microbe-feeding nematodes were removed, the compensatory growth reduced. As a result, while top predators lowered the number of microorganisms feeding nematodes, microbial biomass remained unchanged. Aside from the grazing effect, another obstacle to top-down regulation in soil ecosystems is extensive omnivory, which dampens top-down impacts by increasing the number of trophic interactions. Many organisms can become dormant to resist challenging times in the soil environment, which is a matrix of varying temperatures, moistures, and nutrient levels. Predators and potential prey may be separated by an overwhelming amount of space and time, depending on the circumstances. Because soil food webs are donor controlled, any top-down effects that do occur will be weak. Consumers have little or no influence over the renewal or input of their resources due to donor control. Aboveground herbivores, for example, can overgraze an area and reduce grass populations, while decomposers have no direct influence on the pace at which plant litter falls. They can only impact the rate of input into their system indirectly through nutrient recycling, which, by assisting plants in their growth, results in more trash and detritus falling. Bacterivores and fungivores, on the other hand, would never have a significant impact on the bacteria and fungi they ingest if the entire soil food chain was fully donor controlled. While bottom-up impacts are undeniably important, many soil ecologists believe that top-down effects can also be substantial at times. When predators or parasites are introduced to the soil, they can have a significant impact on root herbivores, affecting plant fitness indirectly. Heterorhabditis marelatus, a local entomopathogenic nematode, parasitized ghost moth caterpillars in a coastal shrubland food chain, while ghost moth caterpillars ate bush lupine roots. H. marelatus was found to be associated with fewer caterpillars and healthier plants. Furthermore, the researchers discovered that in the absence of entomopathogenic nematodes, bush lupine mortality was high. As a natural opponent of the ghost moth caterpillar, our findings suggested that the nematode protected the plant from injury. The authors also hypothesized that the contact was powerful enough to alter bush lupine population dynamics, which was verified by later experimental work with naturally-growing populations of bush lupine. Top-down control is used in agriculture and is the basis for biological control, which is based on the premise that plants can gain from the use of their herbivore's enemies. While wasps and ladybugs are generally linked with biological management, parasitic nematodes and predatory mites are also used to lower pest populations and protect crop plants in the soil. A thorough understanding of the local soil food web is required to apply biological control agents effectively.

Community matrix models: A community matrix model is a sort of interaction web in which each link in the topological web is described using differential equations. The degree of interactions between groups is calculated using Lotka–Volterra equations, which explain predator-prey interactions, and food web energetics data such as biomass and feeding rate. Community matrix models can also demonstrate how minor changes affect the web's overall stability.

Stability of Food Webs

The question of whether complex or simple food webs are more stable has been highlighted through mathematical modeling in food webs. Soil food webs were thought to be very basic until the last decade, with modest degrees of connectance and omnivory. May's mathematical models projected that complexity destabilized food webs, which sparked these thoughts. May employed community matrices in which species were randomly linked with random interaction strength to demonstrate that local stability declines as complexity (measured as connectance), diversity, and average interaction strength among species increases. The usage of such haphazard community matrices drew a lot of criticism. It was discovered in other areas of ecology that the food webs employed to create these models were overly simplistic and did not reflect the complexity of real ecosystems. It was also obvious that the soil food webs did not follow these predictions. Soil ecologists observed that omnivory was frequent in food webs, and that food chains could be extensive and complicated while still surviving drying, freezing, and fumigation. But why are complicated food webs more stable than simple food webs? Many of the factors that prevent top-down trophic cascades also help to maintain stability. If the connection strengths are weak, complex food webs may be more stable, and soil food webs appear to have many weak interactions and a few strong ones. Because it is impossible for main consumers to overextend their resources, donor-controlled food webs may be intrinsically more stable. The soil's structure also serves as a buffer, keeping organisms apart and preventing severe interactions. Many soil organisms, such as bacteria, can become dormant during times of stress and quickly reproduce once conditions improve, making them resistant to disturbance. The usage of nitrogen-containing inorganic and organic fertilizers, which produce soil acidity, reduces the system's stability.

Interactions not Included in Food Webs

Despite their intricacy, food webs struggle to classify some interactions between organisms in the soil. Litter transformers, mutualists, and ecosystem engineers all have significant impacts on their communities that are neither top-down nor bottom-up in nature. Isopods, for example, are litter transformers that eat dead plants and emit fecal pellets. While this may not appear impressive on the surface, the fecal pellets are moister and higher in nutrients than the surrounding soil, allowing bacteria and fungi to colonize them. The bacteria decompose the fecal pellet, increasing its nutritious content and allowing the isopod to re-ingest the pellets. When isopods eat nutrient-poor litter, bacteria enhance it for them, and isopods who aren't allowed to eat their own feces may perish. Similar to the mutualistic link between microbes and cows, this mutualistic association has been dubbed a "external rumen." Isopods rely on microbes outside their bodies, whereas cows' bacterial symbionts live inside the rumen of their stomach. Earthworms, for example, are ecosystem engineers who change their surroundings and provide shelter for other smaller organisms. Earthworms also boost microbial activity by enhancing soil aeration and moisture, as well as moving litter into the earth, where it can be consumed by other soil fauna. By supplementing nutritionally extremely rare food - decaying wood - fungi create nutritional niches for other creatures. This permits xylophages to grow and influence dead wood, aiding in wood degradation and nutrient cycling on the forest floor. The research assumes that competition and predation are the most important interactions in aboveground and aquatic food webs. While soil food webs are ideally suited to these types of interactions, future research should focus on more complicated interactions like mutualisms and habitat alteration. Soil food webs remain an important tool for describing ecosystems, despite their inability to characterize all interactions. The effects of interactions between species in the soil on decomposition are still being researched. However, much remains unknown about the stability of soil food webs and how they change over time. Understanding how food webs affect crucial attributes like soil fertility requires this information.

MANAGING SOIL HEALTH

To improve soil health, you must first recognize that it is a living system. Soil health (better soil function) is primarily an issue of providing sufficient habitat for the diverse animals that make up the soil food web. This can be accomplished by causing as little disturbance to the soil as possible, growing as many diverse species of plants as possible, keeping living plants in the soil as often as possible, and keeping the soil always covered.

Manage More by Disturbing Soil Less

To the world of soil organisms, tilling the soil is the equivalent of an earthquake, hurricane, tornado, and forest fire all happening at the same time. Physical soil disturbance, such as tillage with a plow, disk, or chisel plow that results in bare or compacted soil, is harmful and disruptive to soil microbes, making it a hostile, rather than a welcoming, environment for them to live and work.

Simply said, tillage is harmful to the soil. Misuse of inputs such as fertilizers and pesticides can also cause chemical and biological disturbances in the soil. What happens when we provide the soil inputs? Without any human interference, soil and all the organisms that live and develop in it have been cycling plant nutrients for a very long time. As a result, earth and plants have developed highly efficient and sophisticated ways of collaborating to maintain mutual sustainability. When we add chemical inputs to the soil, we must first understand and respect the current soil and plant connections, or we risk making the system inefficient, if not outright failing. Plants will not form relationships with soil organisms that assist them absorb water and nutrients if agricultural nutrients are supplied to the soil in excess. The plants are left "high and dry" after the "party" and the synthetic fertilizer is gone, with few to no soil factory workers to help them get water and nutrients for the rest of the growing season. Midway through the growth season, the plants give up vital energy (sugars) to link with microorganisms, when the plant should be investing that energy into blooming and seed development to create a harvestable yield. Excess fertilizer, particularly nitrogen or phosphorus, causes plants to become inefficient as they try to function without the support system that evolved with the soil. We can use the nutrient cycles in the soil to give crop nutrients and allow plants to form important connections with soil organisms by decreasing fertilizer inputs. This allows plants to reach their full potential while also allowing the soil to execute all its desired tasks to their best capacity. We can harness a great engine for biological production if we acknowledge the complex life in the soil and work with it rather than against it (growing crops). Workers in the "dirt factory" can be most productive when they have a pleasant working environment with plenty of energy. The farmer is most profitable when they are at their most productive.

Diversify with Crop Diversity

The effective passage of light energy from the sun is required for a living, functioning soil. Photosynthesis is the process through which green plants convert atmospheric carbon dioxide and water into carbohydrates (starches, sugars, lignin, and cellulose) using chlorophyll to absorb sunlight energy. These carbon compounds store the sun's light energy and serve as the building blocks for plant roots, stems, leaves, and seeds. There are two basic methods for carbon to enter the soil and nourish the soil food web's creatures. The first process entails a relationship between plants and certain bacteria, in which the plant's sugars are released from its roots and exchanged to microbes in exchange for nutrients that support plant growth. The second mechanism involves soil organisms devouring dead plant material like leaves, stems, and roots and releasing carbon into the soil as waste products or when they die and disintegrate. Carbon from the atmosphere is transported into the soil as organic matter in this way. Microorganisms in the soil assist in the digestion of organic materials and releasing plant nutrients. To maintain the diverse range of soil microorganisms that reside in the soil, a variety of plant sugars is required. Different plants must be produced to reach this level of diversity. The fact that food and energy networks and webs are made up of multiple species of plants and animals, rather than just one or two, is crucial to increasing soil health. Variety above ground (plants) equals diversity below ground is a guiding principle (the soil food web). Increasing the diversity of the soil food web by growing a diversified crop rotation is an effective approach to do it. Biodiversity is, in the end, the cornerstone to any agricultural system's success. The potential of any farming system is greatly limited by a lack of biodiversity, and disease and insect issues are exacerbated. A healthy and diverse soil food web promotes nutrition, energy, and water cycling, allowing a soil to reach its maximum potential. Increased diversity in a crop cycle and cover crops improves soil health and function, lowers input costs, and boosts profits.

Grow Living Roots throughout the Year

The soil food web is a complex network of creatures that break down crop leftovers and cycle plant-available nutrients through the soil. Every organism consumes something...or is consumed by something. There are various food sources in the soil that fuel the soil food web, but the sugars excreted by living roots are the best. A rhizosphere, or area of concentrated microbial activity near the root, is maintained by living plants. Because it is where the easiest to eat food is accessible and where peak nutrient and water cycling occurs, the rhizosphere is the most active section of the soil factory. Plant roots emit microbial food to attract and feed microorganisms that give nutrients (and other chemicals) at the root-soil interface for the plant to absorb. Growing long-season crops or a cover crop after a short-season crop nourishes the foundation species of the soil food web as much as possible throughout the growing season, because living roots are the easiest source of food for soil bacteria. Nutrient and water cycling occur at a substantially slower rate when carbon is not accessible from live roots. Because the bacteria involved must do more work, they must often delegate parts of jobs to many other species and carry resources further. Sugar from living plant roots is the initial source of food for soil microbes. They next eat dead plant roots, followed by crop leftovers such as straw, chaff, husks, stalks, flowers, and leaves that are above ground. Finally, they feed on the soil's humic organic materials. Soil microarthropods such as mites, springtails, woodlice, earwigs, beetles, and ants must shred dead plant roots and agricultural leftovers. Crop leftovers must be transferred from the soil surface to living plant roots via multiple creatures in long lines. Before the nutrients locked up in humic organic matter are available to the plant, it must be digested by a variety of microbes. How well the soil food chain is fed determines how healthy the soil is. Providing ample food for soil bacteria makes it easier for them to cycle nutrients that plants require to flourish. Sugars from active plant roots, recently dead plant roots, crop wastes, and soil organic debris all provide food for the soil food web's many and varied members. While the Natural Resources Conservation Service's mission statement is to help people help the land, a farmer's mission statement can be to help soil bacteria support the plants by providing the finest soil habitat possible, including food.

Keep the Soil Covered as Much as Possible

Soil cover conserves moisture, intercepts rainfall to decrease their damaging impact, controls weed development, and offers home for soil food web members who spend some time above ground. Regardless of land usage, this is true (cropland, hay-land, pasture, or range). If you want to improve soil health, you should avoid handling it very it on a regular basis. Growing plants and/or their leftovers should always cover the soil, and it should rarely be seen from above. Soil cover is not something that can be taken for granted. Even in a no-till system, crop harvest processes, amounts of residue produced, and low carbon: nitrogen ratios of some crop residues that enable them to decay quickly might result in a lack of soil cover. Soil cover protects soil aggregates from being harmed by the power of rains falling on them. Even a healthy soil with water-stable aggregates (kept together by biological glues) that can endure rains may not be able to withstand a 'pounding.' When crop residues or growing plants cover water-stable soil aggregates, the pounding force of raindrops protects them from disintegration. Water infiltrates the soil and is available to plant roots when soil aggregates remain intact at the soil surface. Early in the growing season, a mulch of crop remains on the soil surface suppresses weeds, giving the intended crop an advantage. This is especially true with a rolled cover crop, which has the potential to cover the entire soil surface at once. They help keep the soil cool and moist, which is ideal for many organisms that start the decomposition of wastes by shredding it into smaller bits. These "shredders" can accelerate residue decomposition and thus nutrient cycling by up to 25% if they have a favorable residue environment. It can be a delicate balancing act to keep the soil covered while enabling crop leftovers to decay (allowing nutrients to be cycled back into the soil). If producers want to keep the soil covered and fed at the same time, they must carefully consider crop rotation (including any cover crops) and residue management.

Dealing with Surface Sealing

Soils are a valuable nonrenewable resource whose relevance has been underappreciated until recently. Soil sealing with artificial, impenetrable coatings obstructs soil's fundamental environmental, economic, and social functions. Soils provide a variety of services, including the provision of food and materials, the regulation of water, energy, and matter, the provision of habitats for biodiversity, the provision of recreational places, and the maintenance of landscapes with aesthetic and cultural value. More natural land is being sealed by impermeable surfaces as a result of an expanding global population and a shift toward increased urbanization: housing, transportation infrastructure, and industrial and commercial development are being laid down over what was once open ground. The influence of this development varies depending on the location of the sealing, the local circumstances and climate, as well as the population size and socioeconomic characteristics of the area. It is difficult for all countries to deal with the growing growth of sealed surfaces.

Investigating Soil Surface Sealing

Understanding soil water circulation through seals necessitates a thorough understanding of Ksat. It's a requirement for modeling soil water movement. When the soil is saturated, the Ksat is a numerical measure of its ability to transfer water. Darcy formulated the notion in 1859. (Hillel, 1998a).

***Use of X-Rays*:** Since Sir Godfrey Newbold Hounsfield pioneered this technology in 1972, X-ray computed tomography (CT) scanners have become widely available. In medicine, the CT scanner is used to image the human body. CT, on the other hand, has been utilized to examine a variety of materials. The scale of observation or resolution of CT scanners can be used to classify them. For macropore investigation in soils, a high-resolution X-ray computed tomography (HRCT) with a 15-m scale of resolution was compared to a medical CT scanner and demonstrated considerable benefits in detecting mesopores. The HRCT scanner (University of Texas at Austin - Department of Geological Sciences) offers an alternative for soil and water measurements. It generates images in a quick, non-destructive, and non-invasive manner. The CT scanner data can be used for qualitative and quantitative analysis. The HRCT images represent serial slices of a sample in a continuous fashion. Data with a spatial resolution of 14 meters or less is provided. Viewing-oriented cuts across 3-dimensional (3-D) sample volumes is possible with image volume rendering and visualization. Solids and voids may be extracted and measured using image analysis, which can be utilized to assess soil surface sealing.

GIS and Indicators: The absolute area of sealed soils, the rate of sealed soils, such as per 1 km2, or per capita, can all be used as indicators of soil surface sealing. Remote sensing data and GIS technology have shown to be effective in determining these indications in the past. From cadastral data, it appears that the surface of sealed soils can be rapidly determined. The cadastral data, on the other hand, does not contain all sealed surfaces. Because the problem mostly affects urban regions and their environs, full data for those areas can be gathered from base maps at sizes of 1:500 or 1:1,000, but this involves a significant amount of effort, even if the maps are already digital. Large-scale digital maps showing buildings, concrete-covered surfaces, bitumen, and green areas, as well as orthophoto-maps derived from aerial photography or satellite images with pixel sizes of 1 m, are the quickest approach to compute the area of sealed soils. On orthophoto-maps, homogeneous zones in terms of development intensity are first delimited. Test zones are established in each homogeneous area (randomly or systematically). The surface of sealed soils is determined using accurate digital maps for which test zone. The sealed surfaces and numerous indicators are then estimated for homogeneous areas and the entire city using the orthophoto-maps.

Mitigation of Soil Surface Sealing

The effect of soil surface sealing especially in agricultural land can be reduced by use of;

Use of Anionic polyacrylamide (PAM)

"A category of very high molecular weight macromolecules generated by free-radical polymerization of acrylamide with an anionically charged co-monomer, primarily the sodium salt of acrylic acid, sodium acrylate," according to PAM. Horticultural Alliance Inc. (Horticultural Alliance Inc., 2006). PAM's charge density and molecular weight are both variable. A charge density of 0 to 100% along the polymer chain can be achieved by altering the acrylamide/anionic monomer ratio, while the molecular weight is regulated by the individual chemical reaction. The United States Department of Agriculture's Natural Resources Conservation Service (USDA - NRCS) now recommends anionic PAM for preventing irrigation-induced erosion. In the United States, anionic PAM is used for water purification and other purposes. When applied for erosion control at recommended concentrations and rates, PAM has been shown to have no substantial detrimental effects on aquatic macrofauna, soil microbes, or crops. Adsorption and deactivation linked with suspended sediments, humic acids, and other contaminants buffer the effects of PAM on biota. The requirement to utilize PAM that contains less than 0.05 percent acrylamide monomer is an essential environmental factor. Acrylamide monomer is a neurotoxic with carcinogenic potential. When used as intended, PAMs with an acrylamide monomer level of less than 0.05 percent are deemed safe. However, acrylamide monomer might irritate the skin. Physical, chemical, biological, and photochemical interactions in soil cause PAM to decay at a rate of 10% each year. The acrylamide monomer is biodegradable in nature, with a half-life of 10 to 20 hours.

PAM is well-known for its ability to prevent soil erosion. PAM can be used either dry or wet. Approximately 1 * 106 acres of furrow irrigated land has been treated with PAM, according to current estimates. PAM can help to reduce soil sealing, improve infiltration, and reduce runoff and sediment transfer. PAM improves flocculation by attaching long polymer chains to soil particles, stabilizing silt and clay soils. PAM, on the other hand, simply serves to stabilize soil aggregation. It won't help a puddled soil's bad structure. Surface applications of 10-kg ha-1 PAM are mostly responsible for surface sealing. Mixtures of PAM with gypsum ($CaSO_4$ * $2H_2O$) may be effective in improving PAM effectiveness in soils with low exchangeable cations. In some soils, PAM has been shown to increase infiltration into furrows. Soil particles are normally suspended in the absence of PAM, eventually filling up the pores and sealing the soil. Infiltration rates are reduced as a result of this filling-in. PAM induces soil particle binding, which reduces particle dispersion and causes soil to flocculate. A 20-kg ha-1 PAM application has been shown to maintain a favorable infiltration rate. It's worth noting that when PAM is applied at high rates, polymer chains might block conducting pores and limit penetration.

Dealing with Soil Crusting

Soil crusts are changes in the top soil caused by natural processes like raindrop impact and the drying process that follows. The effects of soil crusts on seedling emergence and water infiltration are the most significant agronomic drawbacks of soil crusts. Many cultures care about seedling emergence, especially those with little seeds or when the time of emerging is significant for climatic or marketing reasons. In some circumstances, costly replanting is required. Reduced water infiltration in crusted soils causes major irrigation challenges, especially where water scarcity makes large-scale irrigation impractical, forcing farmers to adopt a rational consumption policy. Furthermore, less water infiltration increases surface run-off, generating erosion issues as well as environmental threats due to pollution of surface waterways as a result of possible nutrient losses from increased run-off. Reduced gaseous exchanges between the soil and the atmosphere also contribute to lower crop yields.

Soil Crusting Investigation

Observations of crust morphology, which can range from the field or plot scale using the unaided eye or a low power lens to the micrometer scale using optical and scanning electron microscopes, make a significant contribution to the study of crust formation and problems related to the presence of a crust. Each scale of observation provides unique insight into mechanisms critical to the production of surface crusts. The scale of observation utilized to define crust morphology varies depending on the study's goal, the equipment available, and the researcher's skills, but a thorough knowledge of crust development necessitates the integration of data at all stages. In fact, macro-scale observations are often ineffective at resolving microfabric change, microfeatures, and porosity within a crust, which reveal critical processes in its formation. Observations at a greater magnification using an optical or electron microscope are required to describe these features. Changes in the amount, shape, size, and arrangement of pores, changes in the ratios of groundmass (coarse particles) to micro-mass (fine particles), particle orientation, micro-layering, and other microfeatures associated with crust formation processes have all been observed using optical or electron microscopes. Optical (petrographic) microscope observations were done on thin sections generated from undisturbed samples taken from the surface of a crusting soil using the methods employed in soil micromorphology for thin section preparation. The use of optical and electron microscopes, as well as image processing, allowed researchers to quantify and characterize porosity in soil crusts, which is crucial for water infiltration and seedling emergence. The time and specialized equipment required for thin section preparation and observation, however, limits the technique's use for crust morphology descriptions. The number and intensity of precipitation, as well as the pedological features of soils, all play a role in surface crusting. The employment of a rainfall simulator, which allows for the establishment of the rainfall rate, drop size, and kinetic energy, is extremely ideal for a full investigation of this phenomenon in connection to regional variability and evolution. The study of the links between crust development, surface run-off, and soil erosion requires this instrument. The measurement of crust resistance to penetration is useful for studying the deleterious effects of soil crusts on seedling emergence. The pocket penetrometer can be used for this. It is critical to assess the water infiltration rate, soil sodicity and electrolyte concentration, and organic polymer concentration in order to determine the primary chemical-physical features of the crusts.

Managing Soil Crusting

Tillage has the greatest impact on soil crusting of all soil management methods. Seedbed preparation and weed control are the two main goals of soil tillage. Traditional tillage plays an important role in achieving good tilth, which is a crumbly to granular structure typical of ploughed soils. Such a soil structure is not stable; in fact, long-term intensive cultivations reduce organic matter content, partly because, in modern agriculture, organic materials are applied to the soil at a much lower rate due to a serious lack of farmyard manure. There is a substantial correlation between a decrease in organic matter content and a decrease in the population of living organisms. As a result, the soil aggregate stability drops dramatically, and in crust-prone soils, a single rainfall event is enough to produce the surface crust. When the soil is bare and the surface aggregates are exposed to the direct action of raindrop impact, such as during seedbed preparation and during seedling emergence, the crust forms. As previously stated, crust formation is mostly caused by constant conventional tillage. Conservation tillage strategies like as zero tillage, minimum tillage, surface mulching, contour ploughing, and others have been shown in numerous studies to reduce run-off, soil loss, and are best suited to preventing and controlling crusting. The production of surface soil crusts in a clay loam soil under zero tillage versus conventional tillage was significantly reduced in no-tilled soil compared to traditionally tilled soil.

Zero tillage: No-till farming (also known as zero tillage or direct drilling) is a method of growing crops or grazing without disturbing the soil through tillage. In some soils, no-till farming reduces the amount of soil erosion caused by tillage, particularly in sandy and dry soils on sloping terrain. Increased infiltration of water into the soil, organic matter retention in the soil, and nutrient cycling are all possible benefits. These techniques have the potential to increase the volume and diversity of life in and on the soil. While herbicides are used in traditional no-tillage systems to control weeds, organic systems use a combination of tactics to reduce weeds, such as sowing cover crops as mulch. There are three fundamental no-till farming systems. When crops are sown with seeding machines into a sod created by spraying herbicides to a cover crop, this is referred to as "sod seeding" (killing that vegetation). When crops are sown directly through the residue of a previous crop, this is known as "direct seeding." When crops are left on the surface of the soil, this is known as "surface seeding" or "direct seeding," and it requires no machinery and little effort on flatlands. Today's agriculture is dominated by tillage, yet no-till technologies may be successful in specific situations. Minimum tillage or "low-till" approaches combine till and no-till methods in some circumstances. Some methods, for example, may employ shallow cultivation (e.g., with a disc harrow) but no plowing or strip tillage.

Minimum tillage: Minimum tillage, like strip-till, is a soil conservation strategy with the purpose of using the least amount of soil disturbance possible for successful crop production. It is a type of tillage that does not turn the soil over, as opposed to intense tillage, which uses ploughs to change the soil structure. Primary tillage is fully avoided with minimum tillage, and only secondary tillage is used to a limited level. Practices like as minimum furrowing, organic fertilizer use, biological pest control, and chemical use are all examples of minimum tillage.

Surface mulching: A mulch is a coating of material that is put to the soil's surface. Mulch is used for a variety of reasons, including preserving soil moisture, boosting soil fertility and health, limiting weed growth, and improving the aesthetic appeal of the area. A mulch is frequently, but not always, made of organic materials. It could be permanent (e.g., plastic sheeting) or temporary (e.g., a tarpaulin) (e.g., bark chips). It can be used on bare soil as well as around existing plants. The action of worms and other creatures will naturally mix manure or compost mulches into the soil. When utilized effectively, the method can boost soil productivity in both commercial crop production and gardening.

Contour ploughing: The farming practice of plowing and/or planting across a slope following its elevation contour lines is known as contour bunding, contour farming, or contour ploughing. During times of heavy precipitation, these contour lines create a water break, reducing the formation of rills and gullies by giving the water more time to settle into the soil. In contour plowing, the ruts created by the plow run perpendicular to the slopes rather than parallel to them, resulting in flat furrows that curve around the land. Tillage erosion is also prevented using this strategy. Tillage erosion is the movement and erosion of soil caused by tilling a particular plot of land. Contour bunding, where stones are laid around the contours of slopes, is a comparable procedure. Contour ploughing aids in soil erosion control. Soil erosion control methods like these can dramatically reduce negative effects like reduced crop output, deteriorated water quality, lower effective reservoir water levels, flooding, and habitat damage. Contour farming is a type of sustainable agriculture that is active.

Rotary Hoeing: For shattering crusted soils and allowing seedlings to sprout, rotary hoes can be useful. When deciding on the use of a rotary hoe, consider emerging populations, seedlings, and soil conditions. Because rotary hoes can harm seedlings, the loss from their use (which is typically 1 to 2 percent in maize) should be monitored to see if the damage is larger than planned. Before using the rotary hoe, place various colored flags at emerged and non-emerged seedlings in a defined area to check for potential rotary hoe damage. Following your transit through the area, observe to establish the extent of any damage. Because seedlings are more bendable in the afternoon than in the morning, using the hoe in the afternoon may lessen the danger of harming them. In the afternoon, however, the soil surface may be more difficult to fracture than in the morning. The soil should be damp to dry while utilizing a rotary hoe. If the soil is excessively damp, the hoe is more likely to just punch holes in it rather than fracture it, preventing seedlings from emerging. A smaller tractor can be used to pull the hoe because it usually does not require much horsepower and reduces the risk of compaction. This is also a cost-effective way to prevent soil crusting. The cost of diesel for rotary hoeing is 0.20 gallons (0.75 liters) per acre. Tractor speeds should be between 8 and 10 miles per hour (10 and 20 kilometers per hour), as greater speeds tend to cause fracture. The hoe should only go deep enough to break the crust; any further tilling raises the risk of harming the seedlings. Because driving over the rows and seedlings increases the chance of compaction around the seedlings, the operation should be finished with the rows. Following the hoeing of a strip (perhaps one designated with flags), an assessment of the operation's success should be made. If the hoe looks to be digging too deep and ripping out seedlings, lessen the down pressure or weight on the hoe. If the crust isn't being fractured and the soil isn't excessively damp, increase the down pressure or weight on the hoe. If soil types vary greatly within a field, observations should be performed across them. Hoeing may result in a 3 to 5% loss of plants, which may necessitate moving at a slower speed. Adjustments should be made if the process is knocking off soybean cotyledons, as lost cotyledons usually mean the seedling will perish. Although seedling corn is more forgiving, if considerable damage or uprooting occurs, stopping or changes should be considered.

Dealing with Soil Compaction

Compaction generated by farm operations has been demonstrated to have a variety of effects on field crop output.

- Compaction compacts the earth, making it difficult for roots to penetrate. When roots reach this zone of compaction, they tend to grow laterally. The volume of soil that the roots can investigate for water and nutrients is severely limited as a result.
- Compacted soil also restricts water circulation into the soil, resulting in conditions that encourage soil erosion and runoff. The amount of water that enters the soil and is made accessible for crop use is limited since much of it runs off or evaporates in the case of irrigated lands.
- Because of the destruction of medium and large pores by compaction, there will be less water accessible in the soil for crop usage due to the abundance of micropores that are firmly holding the water.

The amount of yield lost due to compaction varies by crop variety, growing environment, and any practices used to mitigate compaction. Some crops are known to be better at penetrating hardpan than others. Root limitation by hardpan may be less harmful if the crop is often irrigated or receives frequent rains.

Measuring Soil Compaction

It is critical for agricultural producers to maintain their fields in such a way that compaction is minimized. Specific actions must be carried out in fields where soil compaction exists in order to alleviate the compaction. This will ensure that yields are higher and more consistent across the field. The first step in managing soil compaction is to determine whether there is a problem. Using a shovel to dig out a patch of soil together with crop roots is a straightforward approach to observe compaction. Another simple test is to pick out well-established pigweeds or kochia weeds and assess the effect of compaction on the roots. You'll be able to tell whether roots are being confined at certain depths or if they can grow deeper into the soil. A compacted soil layer feels denser than the layers above it, and its structure appears to be in layers rather than fragments. A soil penetrometer is a more quantitative approach to detect a compacted layer (compaction meter). Penetrometers measure the force required to push through soil using a metal probe with a cone-shaped tip. Megapascals (MPa) or pounds per square inch (psi) are a unit of measurement for this force.

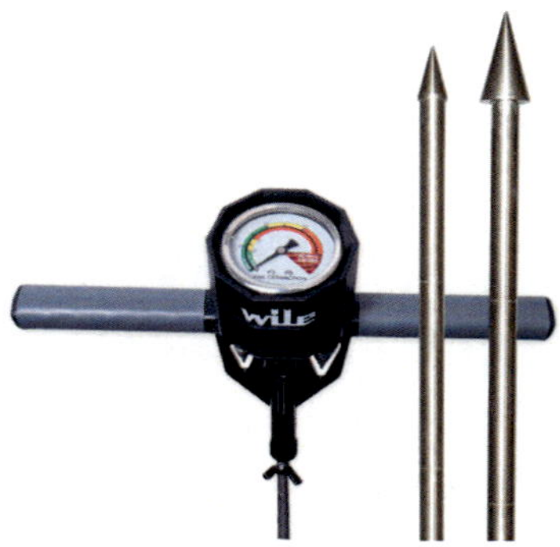

Figure 11: Penetrometer Handy soil compaction tester

Source: Google Images

Most agricultural roots cannot penetrate soil with a pressure of more than 300 psi (2 MPa) (Aase et al., 2001). Because compaction readings rise as the soil dries out, it's critical to take compaction measures while the soil is at field capacity, i.e., after gravity has drained excess soil water from recent irrigation or precipitation. After a heavy rainfall or irrigation event that thoroughly soaked the soil profile, sandy soils attain field capacity in about two days. You may need to wait up to four days in heavier soils before the soil reaches field capacity. When using a penetrometer, it's critical to keep track of the amount of resistance, the depth at which the compacted layer begins within the soil, and the depth at which you break through the compacted layer. You may be able to break through the layer in certain circumstances, while the compacted layer may continue into the subsoil in others. Another precaution to take when using a penetrometer is to take multiple measurements in a specific field to be able to accurately assess the depth of the compaction. Take at least 20 measurements within a uniform acre and determine the average of the measurements. If different portions of the field have different textures or have been handled differently, you can take individual measurements for them. The idea is to take measurements in such a way that you can see where the compacted layer is in the soil for a specific field.

Compaction Management

Whether the compaction is at the surface or beneath the soil, managing a compacted soil layer will be different.

Reducing Surface Compaction

Because the aggregates at the soil surface are weak and sheltered from external forces such as rainfall or flowing water, surface compaction or crusting is common. Increasing the organic matter content of the soil will help to minimize surface crusting and make the soil aggregates more resistant to the effect of rain or running water. Shallow cultivation to break the crust will assist crop emerging in circumstances where a crop has been sown and the crusted surface will hamper emergence. Even after crop emergence, a rotary hoe can be employed to break surface crusts. In some circumstances, softening the soil with irrigation water might help seedlings emerge more quickly. Additionally, mulching the soil surface to prevent quick drying will aid in the delay or prevention of crust formation during the seedling emergence period. Standing stubble can restrict the amount of wind that reaches the soil surface and slow the drying process, reducing the occurrence of surface crusts.

Avoiding Subsurface Compaction

Due to current farming activities that rely on heavy machinery, implements, and tractors, subsurface compaction is a more common type of compaction. Even yet, there are a few techniques to avoid soil compaction in these situations. The following are some better management strategies for preventing subsurface soil compaction:

Avoid soil tillage operations when the soil is wet: Tilling when the soil is wet causes the soil structure to break down more quickly. Soil particles become reconsolidated and positioned closer to each other when soil structure breaks down, forming a compaction layer. The movement of roots and water in the soil will be restricted by this layer. Depending on the soil type, the earliest time to enter into the field after a rainfall or irrigation event varies. After rain or irrigation, fields with sandy soils that drain quickly can be re-entered without destroying soil structure faster than fields with heavy soils that drain slowly. When the soil water is close to field capacity or wetter, the soil structural disintegration owing to applied force is greatest. To avoid compaction, it's best to stay off the field when the soil is quite damp.

Practice controlled traffic on the farm: It is preferable to create permanent farm traffic lanes. Tractors and other farm machinery driven indiscriminately across the field can compact several parts of the field. Establishing movement paths needs forethought and discipline, but it will eventually result in reduced field compaction. Controlling traffic has been demonstrated to increase agricultural soil productivity. To determine where traffic pathways are located, you can either physically mark the field or utilize GPS guidance devices, depending on your agricultural operation. Planting, spraying, and harvesting can all be done with farm traffic following designated paths.

Use tractors with the correct type of tires and tire pressures: Tractor tires disperse forces into the earth, potentially causing compaction. To avoid field compaction, it is critical to use the correct types of tires and pump them to the proper pressures. For spreading out the tractor's weight, larger tires are preferable to smaller, skinny tires. In the field, dual tires have been demonstrated to have a smaller compaction effect than solo tires. Extra tires on duals and triples minimize weight on individual tires compared to single-track tractors since tractor weight is distributed throughout the tires. In addition, the carrying load on each tire is lowered in dual and triple tires, lowering the air pressure required for each tire. The depth and intensity of the compaction are reduced in duals and triples due to the lower tire pressure. It's also a good idea to check the tire pressure on a regular basis to ensure that it's at the optimum level. It's been proven that inflating tires to higher than recommended pressures cause compaction and reduces tire traction. The contour of the tire changes as the inflation pressure drops, with the lower pressure resulting in a bigger footprint and less compaction. Soil compaction was increased about one foot below the soil surface due to the increased inflation pressure.

Reduced tillage practices: Reduced tillage measures, which reduce farm traffic and/or the number of passes, can help to reduce soil compaction in agricultural areas. Soil quality improves over time, and soil organic matter builds up, reducing the impact of soil compaction in agricultural areas, because the soil does not get the entire spectrum of tillage operations as in conventional practice.

Managing the soil organic matter: Organic matter increases the soil's resistance to compaction by making it more elastic, limiting the ability of an applied weight to compact it. Organic matter also contributes to the formation of better and stronger soil aggregates, which can assist the soil tolerate compaction. When compared to a similar soil with low organic matter, soils with high organic matter have higher bearing capacity and can handle agricultural traffic with less compaction. As a result, soil management measures that increase organic matter in the soil will make it more resistant to surface and subsurface compactions.

Mitigating Subsurface Compaction

Breaking subsurface compaction layer with a subsoiler: Subsurface compaction can stifle root growth, preventing crops from growing and yielding properly. As a result, in many circumstances where compaction develops at a fair depth, a deep tillage instrument such as a subsoiler or interrow ripper can be used to break up the compaction, allowing crop roots to grow deeper into the soil. Subsoiling also improves the soil's infiltration rate, enabling more water into the soil profile. It's crucial to determine the depth and extent of the compaction layer before subsoiling. The subsoiler's shank is usually set to a depth just below the compaction layer. If the compaction layer is between 9 and 12 inches thick, the subsoiler can be set to about 13 to 14 inches to break it up. Going deeper with the shanks than necessary wastes energy because subsoiling is an energy-intensive process. In addition, especially in heavy soils, an appropriately sized tractor should be utilized to pull the subsoiler.

Figure 12: A subsoiler

Source: Google Images

Using deep-rooted cover crops: Some crops can penetrate the soil's dense layers. Annual or perennial crops could be used. Such crops can be planted to alleviate compaction in circumstances where the land is not required to be utilised immediately. Cover crops have been demonstrated to minimize compacted soil penetration resistance and bulk density. Crop-based mitigation solutions have the extra benefit of adding organic matter to the soil, which improves soil structure while also breaking up the compaction layer. Forage radish, annual ryegrass, canola, sunflower, sorghum-Sudan, and turnip are examples of crops that can be utilized to loosen up compacted soil layers. Alternating shallow-rooted crops after deeper-rooted crops, or rotating annuals with perennials (for example, alfalfa after annual vegetables), is another strategy for minimizing soil compaction and promoting soil health.

Dealing with Soil Waterlogging

When the soil is saturated with water, waterlogging occurs. Waterlogging is particularly likely in heavy soils. They have a little pore space; thus, water and air can only travel extremely slowly through them. When soil is saturated for an extended period, oxygen is depleted. The roots then stop developing and collecting nutrients, the stomata close, blocking photosynthesis, and denitrification begins in the soil. Plants must remove nitrogen from older leaves to support the formation of new leaves because they cannot absorb it from the soil. During the extraction process, aged leaves become nitrogen-deficient and yellow due to waterlogging. Tillering does not occur because there is insufficient nitrogen in old leaves to support fresh tiller growth.

Measuring Waterlogging

With shallow wells, measure waterlogging in the root zone of the plant. Water perched on subsoil clays gradually saturates the topsoil from below, preventing oxygen, carbon dioxide, and ethylene exchange with the atmosphere. Follow the instructions below to install the wells:

- Drill holes just into the clay subsoil, or to a depth of 70cm if no clay has been encountered.
- Fit slotted PVC pipe to the depth of the hole and leave a short amount above ground if the well will be utilized for the season to assist prevent surface water from entering.
- Fill the slotted PVC pipe with coarse sand to allow water from the clay subsoil to enter through the slots. To keep surface water out, use clay from the hole to make a seal around the PVC pipe's neck.

Managing water-logging

Proper installation and management of surface drainage (including raised beds) is crucial in limiting off-site impacts in order to prevent waterlogging, particularly when sediments and nutrients may infiltrate streams and jeopardize water quality. Drainage alternatives, water storage, and compaction reduction are the greatest practices for mitigating the impact of water logging.

Removal of excess water through drainage options

- To rehabilitate waterlogged terrain and enhance soil structure, surface and sub-surface drainage is often used.
- Over 80% of dairy land now has some sort of surface drainage, with up to 20% having sub-surface drainage (MacEwan 1998)

Surface and subsurface drainage are the two types of drainage systems. Surface drainage is particularly beneficial for transporting surplus water from land to an artificial drainage system or a natural watercourse in a regulated and timely manner. This should be done in a way that does not harm the environment.

Types of surface drainage include:

- *Ditches or open drains*: Spinner cuts or excavators can be used to create these, which vary in size and length. Open drains in dispersive soil should be avoided at all costs since they are prone to erosion.
- *Grassed Waterways*: These are usually shallow, with widths ranging from thin to meters, and are built in such a way that they are frequently grazed as part of the paddock. They're sometimes utilized to divert drain outflows down slopes to prevent erosion without incurring a significant cost.
- *Humps and hollows (bedding):* The practice of shaping the ground surface into parallel convex (humps) surfaces separated by hollows (typically while repairing pastures) is known as hump and hollowing. The humped structure swiftly drains excess moisture, while the hollows operate as shallow surface drains. Humps and hollows are beneficial in places or soil types where tile or mole drainage is not possible.

After you've taken care of the surface drainage, you may move on to the subsurface drainage. It's possible that you'll need to improve the drainage through the soil profile. The goal of subsurface drainage is to remove just the excess water in the soil. As a result, you must first determine the soil type before doing any operation. Types of Subsurface drainage include:

Mole Drains: Mole drains are unlined channels created by pushing a ripper blade (or leg) through the subsoil with a cylindrical foot (or torpedo) connected to the bottom. To help condense the channel wall, a plug (or expander) is frequently utilized. The foot is frequently pointed like a chisel. Mole drains are utilized in heavy soils where ground water cannot flow downward due to a clay subsoil at the moling depth (400 to 600 cm). Mole drains don't drain groundwater; instead, they remove water as it enters from the surface.

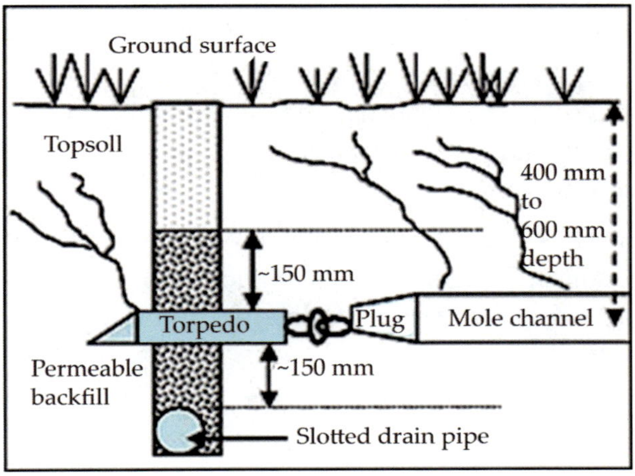

Figure 13: Mole drains over a collector pipe system

Source: David, (2002)

***Gravel mole ploughs*:** A hopper is included onto gravel mole ploughs to allow finely graded gravel to fall into the mole channel. In the United Kingdom, these ploughs have been employed successfully in heavy soils that cannot support "normal" mole drains. Experiments in north-east Victoria and Gippsland have shown that they have promise on unstable clay soils, but they are costly due to the amount of gravel and close spacing required. Unfortunately, in southern Australia, there are extremely few of these machines.

Raised bed cropping: The mechanisms that lead to waterlogging and soil structure deterioration under broadacre cropping regimes have been the subject of intense investigation over the last decade. Raised bed systems, which now cover around 10% of the annual crop area in the Corangamite region, have seen the most advancement. Raised beds are designed to prevent machinery compaction and waterlogging by elevating the soil above the saturated zone and use controlled traffic. Raised beds have enhanced soil structure and reduced waterlogging on cropland where they have been implemented, while also enhancing agricultural yield in heavy rainfall locations.

Minimize compaction - non-drainage options

Controlled Traffic: to improve soil structure and prevent soil compaction

Stock Management: The following methods can be used to manage stock:

- Change the way you use the land (dedicate as a hay or silage paddock and graze only in summer, or remove from the grazing rotation)
- When pugging is impending, remove stock as quickly as possible.
- Use 'on-off' grazing tactics to allocate brief grazing intervals on a limited area to ensure optimal feed intake prior to the onset of pasture damage:
  - After a brief period of grazing, cows are removed from the pasture.
  - Because it maintains healthy ground cover and greater organic carbon levels, it has been found as an effective technique of decreasing hoof compaction on broadacre grazing land.
- Designate a "sacrifice location" where cows will be transported if it rains.
- Build or designate a "loafing area" (pad, laneway, barn, or woodlot) where stock can be moved in wet weather.
- In damp conditions, build a feed pad for any additional feeding.

Land Class Fencing

- Paddocks are fenced based on the land's capabilities and soil type.
- It's easier to keep cattle out of rainy places if paddock fences are placed such that the drier soils and locations are separated from the wetter flat areas.
- Consider land class fence as part of your overall farm plan.

Improve water storage in profile

The likelihood of waterlogging will be reduced by increasing the soil's water holding capacity. Rain will get through if the surface structure is good and there is no surface crust. The structure of the topsoil is improved by:

- increasing soil organic matter levels
- reducing tillage

- increase amount of surface cover to reduce the amount of surface sealing on the soil – this also assists in reduction of water loss through evaporation
- use of raised beds
- adding gypsum

Preventing Waterlogging in Irrigated Agriculture

Irrigation without sufficient drainage can occasionally cause poor drainage by increasing the overall amount of water contained in a soil profile. This is accomplished by reducing excessive water intake into the soil in the following way:

Control of canal seepage: Canal seepage is a major source of water loss and irrigated land waterlogging. It can be managed using the following methods:

- A very essential preventive method is to line canals with impermeable material, such as concrete, clay, or polythene, to control excessive seepage from unlined canals.
- Lowering the canal's designed full supply level in order to reduce seepage losses.
- Increasing the canal's gradient allows for faster flows and hence less ponding.
- Interception drains 15 to 30 meters from the canal banks are excellent at preventing seepage from the canals.
- Convert water conveyance systems from canals to piped systems if at all practicable.
- Controlling canal discharge: During non-irrigation periods, canal discharge can be controlled to reduce seepage losses. Overuse of canal water can be reduced if canal supply is coordinated with groundwater pumping.

In-field water management: This is likely the most critical part of preventing superfluous water from percolating into irrigated lands. It entails the following:

- Applying only the necessary amount of irrigation water (without exceeding the field capacity) so that all of the water is consumed by the plants.
- Drip irrigation, for example, is a more effective irrigation method.

- Convert from surface irrigation to sprinkler or drip irrigation whenever possible.
- Reduce irrigation to a portion of the total command area in locations with high water tables to obtain supplemental irrigation during the dry season.
- Water conservation measures include leveling the fields and not applying more water than the soil can retain.

Control of inflow from other water bodies: Overflowing rivers, flood plains, and channel obstruction can cause flood water to overrun irrigated areas. Flooding that reaches low-lying crops should be avoided: This can be accomplished by"

- Construction of dykes to train the river movement,
- On- stream dams, which act like regulating dams
- Off-stream dams,
- To conserve the catchment with run off collection and disposal systems.

Control of ground-water inflows: Most of the interflow water comes from sloping terrain rivers, with a high penetration rate and the potential for increasing waterlogging. These are governed by the following:

- Construction of an interceptor,
- Growing trees e.g., eucalyptus or crops) that have high water consumption rates (e.g., sugarcane, arrow roots).
- Construction of spring protection works

Control of water from higher ground into lower fields: Terrain water recharge can sometimes inundate farmed lowlands due to excessive water inputs from higher ground. The following would be a good solution:

- Obtain the water for irrigation from somewhere else,
- Proper irrigation water management, and
- Incorporate fish or duck farming.

Preventative Measures in Rainfed Agriculture

Excessive rainfall, flat topography, or inadequate land and water management are all common causes of temporary waterlogging in rainfed agriculture. Preventative methods can be used to control poor drainage, such as:

Reducing unnecessary percolation of rainfall: This entails disposing of storm water in excess of crop water requirements in an efficient manner. Excess water collects due to high seasonal rainfall (known as in-situ rainfall), which is concentrated over a short period and occurs across a large area of flat land, or downstream channel obstruction, or shallow impermeable barrier soils (planosols), or low infiltration rates (vertisols). The best solution would be to establish a;

- Cambered beds,
- Leveling of flat areas to reduce depressions,
- Sub-soiling to break the hard pans and improve infiltrations
- Excavation of farm ponds to store water for later and use/ supplementary irrigation,
- Grow water loving plants e.g., sugarcane, and
- Improving terrace outlets before the rains. A functional outlet is the most important part of the drainage system.

Controlling surface runoff: Surface runoff from large catchments or paved places like homesteads, cities, and highways can be excessive, creating waterlogging in receiving lowlands. High-intensity storms on sloping ground, a high catchment-to-receiving-area ratio, or poor land use and management practices in the catchment can all contribute to excess water. The following would be a good solution:

- Construct diversion ditches/cutoff drains to intercept the runoff,
- Conservation in the catchment areas to reduce runoff accumulation,
- Water harvesting and storage in tanks, pans, and other structures,
- Developing a functional runoff collection and disposal system.

Tree planting: Planting trees, particularly those that absorb a lot of water, such as eucalyptus tree types, can help to lower the water table. The trees help to improve the environment and can also be used commercially. Deep-rooted eucalyptus trees have roots that reach up to 3 meters in length, but they transpire water at a high pace, acting as a biological pump. They can handle unwanted drainage water in irrigation schemes as well as temporary waterlogging in rainfed systems this way. However, bio drainage is only efficient if the root zone has sufficient aeration and the roots grow up to the groundwater reservoir, drawing water from the capillary zone.

Other Preventative Measures

This may include;

- **Land grading:** Land grading can help improve surface drainage in flooded soils, especially in irrigated areas.
- **Cropping patterns:** Plants and crops with a high rate of evapotranspiration can be used to create a natural drainage system.
- **Optimum water use:** Digging wells and sinking shallow tube wells can stimulate the usage of surface and ground water in tandem.
- **Better irrigation methods:** Improvements in water applications by just supplying irrigation water at the optimal depth. To reduce percolation losses, sprinkler and drip watering technologies will be used.
- **Punching underground barriers:** To lower the water table, geological formations such as hidden ridges that may be interfering with subsoil flow and causing waterlogging may be punctured.
- **Efficient drainage system:** Storm water and extra irrigation water must be disposed of quickly, which necessitates a well-designed drainage system.
- **Removing obstruction in natural drainage:** By eliminating garbage from waterways and providing appropriate waterways beneath road bridges and drainage crossings, the environment will be improved.

Dealing with Soil Erosion

For millions of years, erosion has molded landscapes. Human changes in land use and cover, on the other hand, have led erosion rates to rise in many parts of the world, resulting in significant land and environmental degradation. Detachment, transport, and deposition are the three primary stages of soil erosion. The amount of soil that is eroded and the wind or water force's ability to move it determine the severity of erosion. Because both detachment and transport require energy (and deposition occurs when energy is no longer available), the ability of soils to erode is determined by their erosivity (the energy of the eroding agent, such as wind or water), as well as their erodibility (the susceptibility of the soil to erosion). While there isn't much we can do about wind and rain, we can try several tried-and-true soil erosion prevention strategies. The kind of soil, geography, climatic conditions, and other factors all influence which technique (or combination of techniques) is best for your land.

Estimating Erosion

Estimating soil erosion is critical for determining the severity of erosion and its impact on land use and management strategies. Soil erosion is linked to four of the parameters determining the NRCS phosphorus index, and thus manure application rates (NM 13). Because of the large number of variables involved, models are frequently used to estimate erosion. The Revised Universal Soil Loss Equation (RUSLE), which is used to quantify water erosion, and the Wind Erosion Equation are the two most often used models for the western United States (WEQ). RUSLE and WEQ are both described by equations that take into consideration a variety of erosion-related parameters to establish average yearly erosion rates. Climate, soil erodibility, surface roughness, field length, vegetation cover, and, in some cases, an erosion control program are all factors to consider. Because some erosion occurs independently of land use or management, the Soil Loss Tolerable Value, or 'T value,' was created as a baseline against which management-induced soil erosion could be compared. The T value is the maximum amount of soil loss per acre that can be tolerated while maintaining long-term productivity. T values differ by soil type and are influenced by factors such as climate, geography, and the amount of time it takes for new soil to form. Contact your local NRCS office for assistance in determining values for soil loss factors and T values for a specific area.

Managing Erosion

Minimize the erosivity of the eroding agent, reduce the soil's susceptibility to erosion, and inhibit particle transport are all important aspects of managing any sort of erosion. The most effective way to manage erosion is to use current Best Management Practices (BMPs) that focus on five important topics. The following is an overview of a variety of erosion-control BMPs.

Residue Management

The use of plant remains is one of the most useful techniques for mitigating erosion. Standing stubble, distributed straw, growing vegetation, and mulch are examples of vegetative cover that has been left on the field as residue.

Soil erosion is less likely in practices that leave residue on the surface than in practices that remove excess residue. As the amount of residue cover grows, the amount of soil lost via erosional processes decreases.

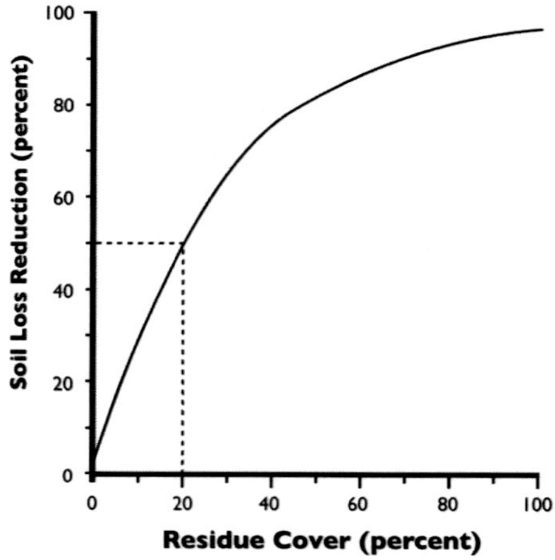

Figure 14: Percent soil loss reduction increases with increasing residue cover

Source: Dickey et al., (1986).

A field with 20% cover, for example, will have a 50% reduction in soil loss when compared to a naked field. For a variety of reasons, residue cover has the effect of minimizing soil loss. Surface residue absorbs part of the energy from the wind, precipitation, and running water, reducing the amount of energy delivered onto the soil. Root systems that are in good shape help to keep soil particles in place. Due to higher infiltration and lower evaporation as a result of less wind and more canopy shadowing, soil moisture is maintained during residue management. Additionally, standing residue will absorb and keep snow better than a bare field throughout the winter months, resulting in increased moisture retention and soil protection. The next sections go over several types of residue management and how they affect soil erosion.

***Tillage*:** Excessive tillage on croplands is a major factor to soil erosion in the northern Great Plains. 'Conventional tillage,' which leaves less than 15% residue cover, burys surface residues, leaving agricultural soils unprotected and vulnerable to erosion. Conservation tillage, on the other hand, leaves more than 30% of crop residue on the surface, providing coverage and protection to the soil surface. Reduced till, ridge till, mulch till, strip till, and no till are all examples of conservation tillage. Conservation tillage frequently improves soil structure by reducing mechanical disturbance and increasing SOM content and aggregation, in addition to providing soil coverage. Minimum tillage (3-5 tillage treatments) produces twice as much surface clod mass as conventional tillage (8 tillage treatments). In the minimum till plots, the results are attributed to decreased clod destruction by tillage machinery and increased residue cover.

***Stubble Height*:** Standing residue reduces wind speed more than residue that is spread out flat on the ground, therefore thus is an important part of residue management. Taller stubble reduces surface wind velocity, provides greater cover, and enhances microsite growth conditions, all of which can improve crop yields. Spring wheat grain output and water use efficiency were both affected by stubble height during the growing season (WUE). When compared to cultivated (disk tilled) stubble, tall stubble (average height 15 inches) boosted yield by 12%, while yield in the short stubble treatment (average height 6 inches) was in the middle. WUE, which was much greater in the tall stubble treatments compared to the farmed stubble, followed a similar pattern. WUE in the short stubble treatment was middle, but not significantly different from WUE in the other treatments. The stubble produced beneficial microclimate growing conditions, such as lower surface temperatures and reduced evapotranspiration losses due to reduced wind velocity on the surface, which were attributed to both yield and WUE outcomes.

Cropping Systems: Crop-fallow systems are widely used to restore soil water and suppress weeds in various parts of the world. However, due to the lack of cover for extended periods, erosion concerns have become more prevalent in many fallow systems. By minimizing or eliminating a fallow time, flexible or annual cropping reduces soil erosion and provides vegetative cover for extended periods of the year. Growing a cover crop during fallow periods as an alternative to crop-fallow provides soil protection and, in the long run, increases water penetration and reduces runoff. Strip farming, or growing crops in strips with fallow in between, helps reduce erosion by reducing wind speed and breaking up the field length, which controls soil avalanching and lowers particle detachment. Strips planted in straight lines and perpendicular to the prevailing wind should be planted to control wind erosion. Strip width will vary depending on the environment, field conditions, and machinery, and can be predicted using WEQ models.

Residue Burning: The practice of burning leftover residue has been around for a long time. Burning allows a farmer to remove surplus residue for easier seeding and plowing, control plant diseases, weeds, and insects to a limited extent, and boost short-term nutrient availability. However, regular, long-term residue burning (>15 years) is thought to exacerbate soil erosion over time due to a decrease in SOM concentration and overall soil quality.

Erosion Control with Vegetation

Shelterbelts: By reducing wind velocity, trapping material, and breaking up field lengths, shelterbelts set perpendicular to the prevailing wind direction can successfully limit erosion. Grass, shrubs, and trees can all be used to make a belt. Climate and moisture conditions, as well as fertilizer and pest management, will all play a role in determining which type and variety of plant to utilize. Perennial grasses, such as tall wheatgrass (*Elytrigia elongata*), have proven to be effective at reducing wind erosion, protecting vegetation, and trapping snow in the general climate. Because of its winter hardiness and drought tolerance, the nitrogen-fixing shrub Caragana (*Caragana arborescens*) is another common shelterbelt species.

Vegetative Filter Strips: Water erosion damage can be mitigated off-site with the use of vegetative filter strips (VFS). By limiting water flow velocity and retaining silt, these strips, also known as buffer zones, minimize off-field transfer and deposition of sediment and other contaminants. VFS have been demonstrated to be effective at reducing sedimentation in runoff water when placed at the bottom of a field or just above the tail water canal in an irrigated system. In general, when the VFS width widens, the sedimentation in runoff reduces. The ideal width is determined by field characteristics like as slope and length, as well as the volume of water moving through the system; steeper slopes and larger fields draining more water would necessitate wider strips. For level fields (1% slope) with a 30:1 drainage area to filter strip ratio, a minimum width of 10 feet is recommended. Furthermore, VFS must be able to support adequate vegetative growth to function during the first significant rain or irrigation event. Fall may be the optimum time to plant a VFS to ensure proper growth.

Erosion Control in Irrigated Systems

Irrigation-induced erosion can be reduced or eliminated by improving irrigation procedures, compacting furrows, keeping crop leftovers in the furrow, or applying amendments. Gypsum and polyacrylimide are two additives used to minimize erosion in irrigated systems (PAM). Gypsum (CaSO4H2O) can help soil particles flocculate (clump together) and improve infiltration, lowering surface runoff and sediment loss. At low concentrations, PAM is a water soluble, organic molecule that aggregates soil and enhances infiltration. It also serves as a flocculating agent, causing dispersed particles in a solution to clump together and settle. This impact causes silt to fall out at the bottom of the furrow and not remain in the runoff water in a PAM-treated system. Anionic (negatively charged) dry powder PAM is the most frequent kind of PAM used in agriculture. Because anionic PAM has a poor solubility in water, it can stay put rather than being dissolved and taken away. Dry powder can be placed as a 'patch' (a 5–6-foot placement of powder on the furrow ditch surface right below the inflow point) or mixed in with irrigation water, which requires vigorous mixing and turbulence to evenly disperse. When sprayed at the NRCS approved rate of 1 lb/ac or 10 parts per million in furrow irrigation systems, PAM has been found to minimize runoff soil loss by an average of 94 percent and enhance water infiltration by an average of 15 percent. Liquid PAM can be used at a rate of 2 to 4 lb/ac in sprinkler systems. To safeguard human and environmental health, PAM requires certain application and handling practices, and PAM products should be purchased in accordance with your state's registration and labeling standards.

Managing Erosion on Grazed Lands

Because pastures are frequently located on sloping territory that is not suitable for crop development, they are prone to excessive soil erosion. Soil compaction and significant surface runoff can be caused by the amount of grazing and intensity of use. Some management strategies, such as preventing overgrazing and finding alternate water sources, can help to reduce soil erosion and animal deterioration. The kind of grazing method, stocking rate, fodder type and growth patterns, and soil characteristics are the key factors influencing erosion on grazed fields. Overgrazing reduces the amount of vegetation that protects the soil from erosion, and increased animal traffic can exacerbate the situation by compacting it. Continuous grazing systems may not be as effective in preventing overgrazing and leaving vegetation intact as rotational grazing systems. Soil stability and structure will be improved by forages with broad root systems. Fine-textured soils are the most sensitive to compaction and should be checked for infiltration on a yearly basis. Compaction in certain locations may be alleviated by amendments and cultivation practices.

Avoid overgrazing: Too many animals in one region for an extended period might result in poor or nonexistent vegetation stands. Soils (especially those in undulating landscapes and near creek banks) are exposed to the erosive impact of surface water runoff when vegetation is sparse. It is far more expensive to reestablish and renovate pasture forage than it is to effectively manage existing vegetation.

Find alternative water supplies: Animals are kept off surface waters and off stream banks by fencing cattle away from waterways and providing clean water via pressured piping systems. Consider nose pumps, solar, wind, or battery pumps, or gravity delivery systems as alternatives to "in stream" irrigation. Getting animals out of the water and limiting their exposure to waterborne microbes may also improve their health and productivity.

Pasture management: Consider grazing in a rotational system. Animals are rotated through a succession of paddocks, allowing for better pasture management and control of cattle access to waterways. Paddocks minimize the concentration of manure (and the nutrients in it), reduce sediment runoff (by minimizing traffic in sensitive areas), and allow vegetation to reestablish in paddocks that aren't in use. You might also consider fencing animals away from creeks, streams, and rivers to create and sustain vegetation in riparian regions (areas adjacent to water bodies). Well-vegetated riparian habitats restrict runoff and allow sediment and nutrients to settle out, reducing bank erosion. Allowing vegetation to establish itself on a creek bank through limited-access grazing can lower water flow, lessen erosive force, allow silt to fall out, and stabilize stream banks.

Controlling Tillage Erosion

Tillage promotes soil loss in upper slope positions, as previously noted. Tillage did not modify the depth of soil profiles at middle slope positions because the dirt in the till layer was only transferred downslope and the removed soil was replaced by soil from upslope. In medium slope positions, when the downslope transport of the till layer is like a conveyor belt from top to bottom of the slope, tillage mostly serves as a transport activity, resulting in zero soil loss and gain. Tillage mixes the soil of the till layer in the short term in toe-slope positions, while the original soil profile is buried under accumulated materials from upslope soils in the long run. Long-term tillage results in shorter soil profiles in convex (or upper) slope locations and deep topsoil accumulations with inverted soil profiles in concave (or lower) slope positions, where subsurface soil material is deposited over original surface strata. Because there is no soil renewal from upslope, long-term tillage removes surface soils in convex (or higher) slope positions, leaving subsoils, parent materials, or bedrock exposed. Farmers in North America and Europe must till into the subsoil; thus, subsoil materials are integrated into the till layer. To compensate for soil losses near the upslope boundary in Asia, farmers must till into parent materials or bedrocks soft enough to be crushed by hoes or plows to obtain a replacement (e.g., Regosols derived from Jurassic Age mudstone), which maintains a certain thickness of the soil layer (generally 20 cm deep). Even though bedrock exposure is not common, a thin soil layer with a high percentage of gravel and coarse texture is always present at the summit positions as a result of parent materials or mudstone pieces being incorporated into the cultivated soil layer. Because of soil erosion caused by tillage, the depth of soil profiles at the slope's summit is usually so small that only the cultivated layer can be seen. This contrasts with the soil profile with multi-genetic horizons (e.g., Ap, AB, B), where an original subterranean horizon is directly exposed at the surface and forms a new till layer. Conversions from conventional to conservation tillage can significantly reduce tillage erosion rates. Contour tillage is a widespread and successful technique for reducing soil loss due to tillage. Contour tillage, which uses a duck-foot chisel, reduces soil translocation by 51% in southeast Spain when compared to standard up- and downslope tillage. Tillage-induced soil loss was also shown to be decreased by almost 70% employing contour ridge tillage on steep slopes in Claveria, Philippines, when compared to moldboard plowing up- and downslope. When contour tillage was performed by hoeing instead of downslope tillage, a comparable reduction (77%) was seen in the Sichuan Basin, China. "The non-overturning tillage," which can reduce tillage erosion rates by 63 percent when compared to conventional tillage, has had a substantial effect in reducing tillage erosion rates in some specific conditions (i.e., turning soil and pulling it downslope).

Conservation Tillage

Conservation tillage's overall purpose is to protect a farmer's soil while also reducing erosion. You may fix your farm's soil erosion problem by implementing appropriate soil preservation measures, as well as progressive land management, tillage, and cropping methods. Conservation tillage procedures maintain the soil's surface and allow water to seep into the ground rather than flow off. These practices are divided into three categories by scientists: no-till, ridge-till, and mulch-till.

No-Till: From harvest until planting, the no-till method leaves the soil relatively undisturbed. Create a tight, 6-inch-wide seedbed with an in-row chisel or rototiller. Plant your crop and then use a press-wheel to ensure that the dirt and seeds are in contact.

Ridge-Till: When your fields are virtually level and have poorly drained soils, ridge-tilling is the best option. The ridges help to speed up drainage by preparing your beds with sweeps, disk openers, or row cleaners. Planting on the contour helps the soil warm up early in the spring and reduces soil loss. This practice also leaves plant detritus on the surface of the soil between rows, delivering a much-needed nutrient injection to the ground. From harvest through planting, however, the land is left undisturbed.

Mulch-Till: Mulch-tilling involves tilling the soil with chisel plows, disks, and sweeps before planting, making the ground rough and cloddy. This method of tillage does not invert the soil, and the number of blades has an impact on the amount of residue cover that remains on the soil's surface. In the fall, chisel to a depth of 8 to 10 inches, whereas in the spring, only chisel to a depth of 6 inches. For seedbed preparation, choose shallow disking. There is no one-size-fits-all solution for every garden or field, and crop rotations and tillage improvements are sometimes insufficient to address erosion issues. A combination of techniques is frequently required, and in extreme circumstances where concentrated runoff occurs, structural controls such as water and sediment basins are required. Many farmers are concerned about soil erosion. Thankfully, conservation tilling is allowing the agricultural industry to make tremendous headway in addressing erosion issues.

Controlling Water erosion

Land depletion by water can be natural or expedited, depending on the source of its development. Natural water erosion is uncontrollable by humans and has no impact on soil fertility. Rainfall, melted snow, or runoff, in other words, natural processes, are to blame. Each soil type has its own natural erosion rate, which is determined by farming characteristics and environment. Irrational farming, on the other hand, has resulted in increased erosion. It occurs when the irrigation method, volume of water, and time of application are all incorrect, resulting in the destruction of the fertile layer of the soil. Crop yields are reduced as a result of soil erosion caused by water. The greatest approach to keep soil from being carried away is to manage your soil and water resources. In most cases, soil erosion is controlled by a system made up of several components rather than a single technique. Each component serves one or more purposes. The functions that a system must perform are as follows:

- control rill and inter-rill erosion,
- deposit eroded material, and
- convey runoff water in a non-erodible manner.

***Conservation Tillage*:** "Any tillage sequence that lowers soil or water loss relative to conventional tillage," according to the definition of conservation tillage. Conservation tillage offers a lot of promise when it comes to minimizing rill and inter-rill erosion. On practically all lands, conservation tillage is an important practice for minimizing erosion. The efficiency of erosion control varies substantially depending on the tillage strategy used and the amount of crop residue available. Erosion reductions might be as low as 0% to as high as 90%. In the United States, conservation tillage is the most common approach for preventing rill and inter-rill erosion. Conservation tillage has been shown to be an effective technique of reducing erosion since 1974, according to economic research. This is a regular occurrence in the northern Corn Belt's chilly, damp soils. Many states have produced tillage guides to help farmers choose the optimal tillage strategy for their soils. Conservation tillage has been demonstrated to significantly reduce rill and inter-rill erosion, although its impact on runoff volumes has been inconsistent. Conservation tillage has reduced runoff volumes by a significant amount in certain cases, whereas the reduction, if any, has been negligible in others. As a result, conservation tillage systems must contain components for transferring surface runoff to channels in a non-destructive way.

***Sod-Based Rotations*:** There is no better way to control rill and inter-rill erosion than to keep land in a continuous meadow. Sod-based rotations are crop rotations that include grassland. Early American agriculturalists supported the use of sod in rotation for soil formation as early as the mid-1700s. Sod-based rotations reduce erosion by temporarily lowering the soil's erodibility after the sod is tilled and reducing the frequency of row crops, reducing the amount of time the soil surface is bare and farmed. A sod-based rotation that includes both a row crop and a minor grain for one-third to one-fourth of the year would be projected to have roughly 20"70 as much erosion as a continuous corn crop and less than 20% of the erosion of a corn-soybean rotation. Estimates based on the universal soil loss equation (USLE) match those measured at the two Iowa sites well. Soil erosion in meadows is quite minimal when using a sod-based rotation. The USLE coefficients demonstrate this (Wischmeier and Smith, 1978). Soil erosion in a high-producing grass and legume meadow is around 20% of corn erosion under no-tillage agriculture. Even the most erodible meadow is just slightly more erodible than maize grown without tillage. Soil erosion is less after a meadow is tilled than after a planted crop. The first year after treatment, the reduction is 40 to 75 percent, and the second year, 5 to 50 percent. The advantages of a sod-based rotation go beyond reducing soil erosion. A well-managed, sod-based rotation can keep soil organic matter at an acceptable level for a long time. When erosion management expenditures are factored into crop production costs, sod-based rotations may be more profitable than continuous row cropping in some situations.

Contour Farming: In an 1813 letter to C. W. Peale, Thomas Jefferson called for "horizontal plowing": Our land is hilly, and we've been in the practice of plowing in straight lines, whether up and down the hill, in oblique lines, or wherever they lead; and our soil was all quickly draining into the rivers. We now plough horizontally on the dead level, following the curvatures of the hills and hollows, no matter how crooked the lines are. Every furrow then serves as a reservoir, receiving and retaining the water, which is subsequently used to benefit the growing plant rather than rushing off into streams. Almost no soil is transported away from a farm that has been ploughed horizontally and deeply. While more contemporary conservationists recognize the significance of contouring, they do not believe it provides as much erosion prevention. In general, contouring is particularly successful in limiting soil erosion during storms with minor volumes of runoff. Contouring, on the other hand, is less successful where runoff rates are high. Contoured rows may be topped on long slopes, concentrating runoff, and generating significant riling or gullying, and soil erosion may be increased as a result of contouring. Contouring clearly loses its usefulness after a certain period. Contouring appears to have a minor impact on soil erosion in gently sloping locations with low erosion. As slopes become steeper, runoff velocities and erosion due to riling rise. As the slope grows, the critical slope length decreases, and contouring becomes useless. Contouring is a cost-effective option for land management because it does not contribute significantly to the costs of farmers. While contouring is a cost-effective land management option, it is frequently insufficient to prevent soil erosion. Furthermore, huge farm machinery is unable to follow contours precisely, resulting in contoured areas of fields and excessive row slopes. There is a limited database for examining the effects of deviations from the contour.

Contour Strip-cropping: The benefits of contouring and sod-based rotations are combined in contour strip-cropping. Strip-cropping is the practice of rotating equal-width strips of various crops, with at least one strip in a close-growing crop or sod. Each year, the crops on each strip are rotated. Contour strip-cropping is fairly like a sod-based rotation in terms of cost, but with certain adjustments to account for contouring. For contour strip-cropping, Wischmeier and Smith (1978) advised maximum slope lengths and maximum strip widths. Another advantage of contour strip-cropping over sod-based rotations or contouring is that soil eroded from row-crop strips is often caught in strips of meadows or close-grown crops, preserving soil close to its point of origin and on the field where it might contribute to future crop production. To avoid soil deterioration, runoff water from strip-cropped regions must be transferred. Conservation tillage, however untested, can be employed instead of conventional tillage in strip-cropping. Such strips could be utilized to lessen the frequency of sod or close-grown crops in strip-cropping in some cases.

Terracing: A terrace is a place where you may sit and relax "a man-made measure primarily intended to decrease runoff in high-rainfall areas and conserve water in low-rainfall areas. Controlling erosion is the goal in more humid places, while it is a critical goal in drier areas ". The alignment, cross section, gradient, and outlet of terraces can all be characterized. Terraces might be parallel, farmed over, have steep front and back slopes, be level or on a grade, and have surface or subsurface exits, both, or none. In addition, the spacing between terraces may be partially or completely level (or nearly so) for optimum crop production usage of available moisture. Terraces reduce erosion by shortening slopes and preventing or minimizing damage from surface runoff. Conservation tillage helps to preserve soil resources by reducing rill and inter-rill erosion between terraces, as well as lowering maintenance costs and extending the life of terraces. Terraces have a significant impact on sediment discharge from a field. Terracing for soil erosion management is not a new practice; it has been done in other nations for millennia. Although there was less attention on having terraces parallel, it was known that farmers disliked acute bends in terraces and that terraces in these locations should be built higher and wider.

Grassed Waterways: Open channels are commonly used to transport runoff water to watershed exits. If left unchecked, these waterways contribute significantly to watershed sediment yields and cause field dissection. Drill and inter-rill erosion on the field may be considerably exacerbated as a result of increasing slope steepness owing to channel degradation. Constricting flow to non-erodible, grassed rivers is one of the most effective techniques to reduce the threat of erosion in open channels. Such channels have a well-established design, with design information available in guidelines and manuals. Information on construction and maintenance costs, as well as sediment trapping, is not publicly available. The quantity of sediment that accumulates in grassed streams has mostly gone unnoticed. It is estimated that grassed rivers trap nearly 70% of the soil entering watersheds in Texas' Blacklands. Waterway establishment and upkeep are challenging and expensive due to significant soil erosion above grassed streams and deposition within waterways. Many grasses in streams are also vulnerable to herbicide harm. Although waterways necessitate a significant amount of agriculture, many fields would be useless without them. Grassed streams have a wide range of initial expenditures, ranging from $10 per hectare of watershed area to much higher sums when tree clearance and earthmoving are involved.

***Underground Outlets*:** Underground outlets can be a more appealing option for transferring runoff water to watershed outlets than grassed rivers. They function similarly to urban storm drains and require meticulous engineering, construction, and maintenance to ensure proper operation at a low cost over time. Surface runoff ponded on a watershed is drained through underground outlets. Ponding can occur above terraces, in field depressions, and above water and sediment management basins and diversions. Peak runoff rates are considerably lowered when runoff is ponded, and significant deposition occurs in the pondage area. Because ponding is only temporary, just a small amount of land is taken out of production. Subterranean outlets must be carefully built to ensure that discharge rates are adequate but not excessive, and that the underground outlet is protected from damage caused by high velocities and excessive water pressure. Also, when numerous ponded areas drain through a single outlet, effective engineering is essential to prevent overtopping of lower ponded areas due to runoff water from higher elevation ponded regions being discharged into lower ponded areas. Subterranean outlets and inlets to underground outlets must be maintained to avoid clogs. If this isn't done, ponded areas become unfarmable, subsurface outlets are damaged, and structures are topped. When structures are included, underground outlets provide for a lot of flexibility in field design. They're frequently essential to achieve the straight, parallel rows that huge farm machinery demands.

***Water and Sediment Control Basins*:** Structures over important waterways produce water and sediment control basins, where surplus runoff is temporarily ponded, usually over standing crops, until it empties through underground outlets. The basins are primarily intended to manage runoff in order to prevent gullying, promote sediment deposition, and improve the efficiency of farm equipment. Water and sediment control basins typically have little impact on rill and inter-rill erosion, which is a key distinction from terraces. Point rows are often avoided by placing them parallel to field boundaries. The impoundment element of CREAMS can be used to quantify the effect of basins on sediment output from a field (Chemicals, Runoff, and Erosion from Agricultural Management Systems). Typically, sediment trapping of more than 80% of the sediment transported to basins is predicted. Water and sediment control basin maintenance expenses vary greatly depending on the watershed. Where conservation is not applied above a basin, deposition could be so great that the storage volume is filled in a short period of time, perhaps less than a decade. When good erosion control is used, such as rotations and conservation tillage, the life of a basin can be extended by a factor of 10 or more without maintenance. Maintenance expenditures could be lowered by a similar proportion. Most of the benefits of water and sediment management basins are in limiting gully growth and off-site sediment damage because they do very little, if anything, to reduce erosion on the watershed above the basins.

***Vegetative Cover*:** Many regions within fields are either unsuitable for cropping or, if farmed, are prone to soil erosion. Unevenly shaped areas, service roads, turn areas at row ends, unproductive areas with physical or chemical issues, and other particularly erodible locations are examples of such areas. Vegetation is a key means of reducing soil erosion, allowing access, and enhancing farming efficiency in most of these places. Both agricultural and non-agricultural areas can benefit from vegetation as sediment filters. Grass filtration strips work wonders for preventing soil loss on building sites. As machinery has grown and land has become more precious, turn areas at row ends have become a growing challenge. Turn sections for 12-row-wide equipment with 0.75-m row width and 800 m (0.5 mile) row lengths would take up around 2.3 percent of the field. As row lengths shorten, the percentages rise. Turn areas are prone to heavy tillage, poor crop development, and are typically up-and-down slope, especially when contouring is used. Compaction from harvesting, grain trucking, and tillage frequently necessitates more tillage. Grass strips in these regions help to alleviate these issues while also providing an excellent transportation area. These grass strips are also degraded material deposition locations. Problem areas with vegetation are no longer a cause of soil erosion. Although establishing grass in soils with significant physical or chemical difficulties can be difficult, some grass species can generally be planted. Vegetation must be carefully monitored to ensure that insects and illnesses do not endanger crop output. Inter-seeding of fields with cover crops, grass, or small grains is an excellent strategy for decreasing soil erosion after the harvest of an earlier crop and before or during the planting of a succeeding crop, in addition to providing vegetative cover for certain regions of fields. There are numerous possibilities. Such a strategy may or may not be commercially successful depending on the farming operation and various other criteria such as climate, preceding and following crops. Soil erosion can be reduced by using a set of simple soil conservation activities. The components must first reduce rill and inter-rill erosion to a safe level, then convey runoff water from fields to watershed outlets and deposit eroded soil before it reaches the transportation system. Before erosion control on a field can be regarded good, special treatment for peculiar regions must be reviewed. In some cases, some options retain the soil resource better than others. In every situation, however, a successful system for sustaining soil resources should be designed to limit erosion to an acceptable level, carry runoff water to a non-erosive exit, and place structures that induce sediment deposition where they are needed.

Controlling Wind Erosion

The greatest way to prevent wind erosion is to have a sufficient surface soil texture. Wind erosion can be efficiently reduced by properly managed crop residues, carefully timed soil tillage, and precisely set crop strips and crop barriers. The key techniques that aid in wind erosion control are proper land use and application of suitable moisture conservation practices. Several cultural practices can aid in the reduction of wind erosion in arid and semiarid locations where the problem is frequent. Soil roughness or soil wetness, in the absence of crop residue, can significantly prevent wind erosion. Controlling wind erosion can be done in three ways:

- Maintain Vegetative Cover (Vegetative Measures)
- Roughen the Soil Surface by Tillage Practices (Tillage Practices or may be called Tillage Measures)

- Mechanical or Structural Measures (Mechanical Measures)

There is no one-size-fits-all approach to erosion control because so many variables influence the outcome. However, by knowing how soil erodes, solutions to reduce erosion can be established. Wind erosion can be reduced by a variety of methods, but they all aim to achieve one or more of the following:

Reduce wind velocity at the soil surface: This is done with windbreaks;

- crop residues,
- cover crops,
- surface roughness, and
- strip-cropping.

Trap soil particles: This is accomplished by;

- maintaining crop residues on the soil surface
- ridging or roughening the soil surface — all to trap moving soil particles.

Increase size of soil aggregates: This is done by;

- using crop rotations that include grasses and legumes,
- growing high-residue crops and returning crop residues to the soil,
- applying manure, and
- by using emergency tillage, which can create stable clods on the soil surface if soil moisture and texture allow. Increasing the size of aggregates means that it takes a stronger wind to move the soil. A protective cover of living vegetation or crop residues on the surface is the simplest and surest way to reduce wind

Unless there is a prolonged drought, in which case crop residues may be difficult to generate and maintain. Wind speed is reduced at the soil surface by crop residues or growing vegetation, which prevents much of the wind force from touching soil particles. Moving soil particles are trapped by vegetative matter on the surface, which minimizes the avalanching impact. The amount of crop residue needed to reduce wind erosion varies depending on the residue type, height, wind direction, and soil type. By leaving crop residues on the soil surface and limiting soil pulverization, conservation tillage lowers wind erosion and reduces soil pulverization, which can occur when soils are tilled when dry. Tillage reduces the amount of surface residue and dries the soil, making it more prone to wind erosion. Tillage may bring erosion-resistant clods to the soil surface if soil moisture is adequate, but this is usually only a temporary solution. The following are some of the most prevalent strategies for preventing wind erosion:

Windbreaks, wind barriers and shelterbelts: Windbreaks, wind barriers, and shelterbelts help to limit wind erosion by lowering the wind speed on their leeward side below the threshold required for soil movement to begin, as well as shortening the field length along the erosive wind direction. The effect of any barrier on slowing soil movement is determined by wind speed and direction, the threshold velocities required to commence soil movement, and the shape, width, height, and porosity of the barrier.

***Annual and perennial grass barriers*:** On dryland cropping regions, perennial grass barriers have proven to be effective at controlling wind erosion, trapping snow, and reducing evaporation. Annual crops are sometimes interplanted in narrow strips or rows to protect the other crop or a fallow land that is not protected.

***Polymeric emulsions*:** These are non-vegetative covers that are mostly, but not entirely, employed on non-agricultural terrain where growing and controlling vegetation is not viable. Polymeric emulsions protect soil from wind erosion by producing a film.

- Surface crust that is water-insoluble, resilient, and porous enough to allow water to percolate.
- weak enough for seedling penetration
- When utilized as permanent wind erosion control covers, they must be able to maintain their adhesive properties continuously.
- must be easy to apply.

***Artificial barriers*:** Snow fencing, board walls, bamboo and willow fences, earthen banks, hand-inserted straw rows, and rock walls have all been employed to combat wind erosion on a limited scale. These barriers are normally quite expensive to build in terms of material and manpower, and their application is mainly limited to high-value crops. They can also be utilized in sand dune areas to help with the initial stabilization of the land while grass and trees grow.

Strip cropping: This is the process of dividing a vast field into thin strips to assist control wind erosion. Strips of erosion-resistant crops or standing stubble are usually alternated with other crops or fallow fields that are not protected. Strip cropping is a particularly effective erosion control approach because it lowers the downwind avalanching effect of wind erosion.

***No tillage*:** A no-till crop is one that is planted straight into a seedbed that has not been tilled since the previous crop's harvest. There is no extra tillage, and only the immediate seed zone is disturbed. Between harvest and the planting of the following crop, reducing the number of tillage operations is an important step toward reducing soil erosion.

***Cover crops*:** The primary goal of a cover crop is to provide vegetation that protects the soil from wind and water erosion. When protective leftovers are insufficient and winter and spring winds are excessive, cover crops are typically grown. They're placed between rows to prevent seedlings from abrasive harm when they're young.

***Roughening the land surface*:** Embedded vegetative material, soil ridges, soil clods, or mixtures of these three make up soil surface roughness. All these factors contribute to wind erosion management by reducing wind velocity at the soil surface and shielding erodible soil fractions.

***Emergency tillage*:** It's a last-ditch wind erosion management method that can result in a slick, cloddy surface. Excessive grazing, dryness, inappropriate or excessive tillage, or producing crops that yield little or no residue are all common causes of decreased vegetative cover. Before blowing begins, emergency tillage should be completed since soils become more erodible as a result of the abrasion of moving particles, necessitating even more stringent actions to prevent additional erosion.

***Geosynthetic erosion control products*:** When a geosynthetic envelopes or surrounds elements like sand, boulders, or straw, it provides confinement.

Table 17: Effect of tillage equipment on crop residue cover

Implement	% Of residue cover remaining after one pass
mouldboard plow	10
chisel plow (less than 12-inch)	50 to 70
sweeps (20-inch to 30-inch)	80
blade (more than 30-inch)	90
offset disc	50
tandem disc	60
harrow-spring-tooth	65
harrow-steel tooth (more than 12-inch)	95

Source: https://www.alberta.ca/wind-erosion-control.aspx

Natural fiber mulches, geotextile filters, erosion control netting, riprap, erosion control blankets, and open weave meshes are just some of the options. Vegetative density, high moisture absorption, flexibility over differential settling, and strong sediment yields are some of the features these goods bring to the soil.

***Protecting summer-fallow fields*:** Tilling summer-fallow fields dries up the topsoil, making it more prone to erosion. Wind erosion is particularly dangerous in bare fields in late winter and early spring, when the fallow period is coming to an end and the following crop is being seeded. Leaving as much crop residue cover as possible during the fallow period is the greatest strategy to conserve soil moisture and minimize wind erosion. Evaporation is reduced by the crop residue layer, while standing stubble stores snow for extra spring soil moisture. Crop residues also diminish wind speed at the soil surface, while standing stubble anchors the soil. Some farmers opt for chem-fallow, which eliminates tillage and relies solely on herbicides to maintain a crop residue cover. Others use herbicides to replace one or more tillage operations on fallow land. In late fall or early spring, for example, you can use 2,4-D to manage winter annual weeds. This permits you to postpone the first cultivation until late May or early June while the land is still fallow. Also, if configured for minimum activity, rotary harrows can manage weeds without disturbing the residue cover. If you do decide to till, keep your tillage speed to no more than 8 km/h (5 miles/h). Use wide blades or low crown sweeps to ensure residual cover when growing. Remove the mounted harrows from the cultivator to leave more standing stubble. After late August, avoid cultivating summer-fallow. The remaining weeds will not drain much more moisture from the soil, but they will aid in the trapping of snow and reducing wind erosion over the winter. When passing over eroded knolls, raise your implement up instead of tilling them.

***Protecting continuously cropped fields*:** In continuous cropping systems, wind erosion is mainly caused by plowing procedures that bury most or all the preceding crop leftovers. Tillage reduction or elimination will maintain a residual cover for erosion management as well as soil moisture conservation. Use a low tillage pace, implements that bury less residue, and avoid tilling eroded knolls if you must till. Crop residue management is required in reduced or zero-tillage systems. Harvesting is the beginning of residue management. Spread chopped straw and chaff uniformly across the whole width of the cut with the combine. Plugging in sowing equipment and uneven crop emergence are also difficulties that can be avoided with uniform spreading. You may need to use oscillating harrows to distribute clumps of straw after harvest. In the spring, a rotary or flail mower can be used to cut through heavy residues or tall standing stubble. Burning an unharvested crop or a thick, poorly scattered straw is not a good idea. Burning removes humus and organic carbon in the soil, making it more prone to erosion. Use high clearance planters for direct seeding systems. Sowing into firmly embedded stubble is often easier than seeding through loosened stubble. Ground openers come in a variety of sizes and styles, from those that hardly disturb the soil to those that disturb it enough to give some weed control while sowing.

Protecting irrigated fields: Residue management on irrigated fields frequently entails burying the heavy crop remains produced by high-yielding crops (for example, soft wheat). As a result, these areas may be vulnerable to wind erosion. The greatest strategy to reduce wind erosion on irrigated land, like with other crop production systems, is to maintain a crop residue cover by limiting or eliminating tillage. If possible, use the combine to cut and evenly scatter crop remains. Make sure the combine's chopping and spreading machinery does a good enough job in the spring to allow the seeder to clean the residues. Mowing very heavy leftovers with a flail or rotary mower is another option. Another alternative is to bale and remove heavy residues, if enough residue remains to protect the soil from erosion. Sugar beets, potatoes, and beans are examples of irrigated crops that leave little residual cover. Wind erosion on certain fields necessitates special treatment. Seed oats or rye as a cover crop for crops that are harvested early. The cover crop has enough time to mature before winter to protect the soil, but it won't cause a residue problem in the spring. Wind erosion is more prevalent in the non-irrigated corners of pivot-irrigation fields. Wind erosion can be prevented by planting forages or grass on these corners.

Protecting forage fields: Wind erosion can also cause harm to forage crops. When fields are torn up with heavy disks and plows to convert them to annual crop production for weed control and stand rejuvenation, the biggest risk exists. The greatest strategy to limit the risk of erosion from forage conversion is to seed forages, cereals, and oilseeds directly into forage stands. After a forage stand has been killed with a herbicide, crops can be straight sown into it. Forage plants are effectively suppressed by glyphosate spraying in the fall. When compared to tillage, research suggests that you can have the same or superior yields.

Protecting pastures: Overgrazing diminishes pasture yield over time and exposes soil to erosion. Grazing must be controlled in order to maintain adequate plant cover. In general, leave around 50% of the present growth for native range and 25% for tame pastures in permanent pastures under continuous grazing.

Dealing With Soil Nutrients Loss

Nutrients are important and useful resources that can help feed productive grazing pastures and crops. From a cost standpoint, it makes logical to match nutrient delivery to grassland and crop requirements, limiting nutrient loss to the greatest extent possible. As a result, the extra expenditures (e.g., tractor fuel, spreading equipment, labor, etc.) incurred when nutrients are applied beyond the crop and grass requirements could be reduced. Furthermore, nutrient loss might result in additional costs for the farmer; for example, to avoid soil acidity (which can be exacerbated by fertilizer use), farmers may need to lime their soils. Avoiding the negative consequences of nutrient loss benefits farming enterprises by preserving soil health and fertility, as well as crop production. The following are some methods for reducing nutrient loss:

Composting manure: Solid manure composting converts nitrogen to ammonia while reducing the volume of the original material. Composting is a simple process that produces a product that can be simply exported off-farm. Manure can be composted in containers, but it needs to be aerated on a regular basis by pushing or dragging air through the heaps. Closed systems and changing the C:N ratio can assist reduce ammonia emissions (up to 40% of N losses are in the form of ammonia), but they can also increase nitrous oxide emissions. Solid pig/cattle manure and pig/cattle/poultry deep litter are the best candidates for composting. Aeration of the compost on a regular basis may necessitate the use of specialized equipment. Moisture content, porosity, C:N ratio of raw manure, temperature, pH, and forced aeration/mechanical turning all affect the quality of compost generated as well as the amount of gaseous nitrogen lost owing to denitrification during composting. Composting is most effective when the C:N ratio is between 20 and 30. Straw must be added to pig dung to achieve the C:N ratio essential for optimal microbial activity. Composting makes it easier to store and use manure. Compost as a fertilizer also improves crop nutrient uptake, lowering the demand for chemical fertilizers. With no odor, pathogens, or weed seeds, composted manure can be kept or dispersed. Furthermore, higher availability of certain N (ammonium) for plants may result in reduced losses to water or air after spreading. When rain or snow falls on exposed compost piles, water percolates through the piles, allowing nutrients to be transferred to water bodies via run-off or leaching. However, more than nitrate run-off and leaching losses, ammonia volatilization may limit compost's effectiveness as a plant nutrition source. The following are some of the environmental consequences:

- ***Air***: Reduced ammonia emissions (depending on the C:N ratio) result in less particulate matter, odor, and ozone production, lowering health concerns (e.g., aggravated asthma and respiratory problems).
- ***Water***: Reduced nutrient leaching and run-off (to a lesser extent N and P) due to greater ammonium content and storage in soil organic matter, reducing eutrophication and enhancing surface and groundwater quality, lowering threats to human health and biodiversity.
- ***Biodiversity***: Reduced nitrogen deposition has a positive impact on natural ecosystems.

Cooling manure or slurry: Controlling the temperature of stored manure or slurry to reduce the rate of volatilization and consequently ammonia emissions. To avoid temperature variations caused by the outside temperature, properly insulate the storage units. Install cooling plates and pipes, and monitor the temperature on a regular basis. This measure necessitates the storage of slurry in a tank. It may necessitate farm-level infrastructure adjustments, such as the placement of slurry tanks beneath dwelling units, which may warm the manure due to its proximity to the cattle. The cooling slurry could also be coupled to a heat recovery system. Reduced fertilizer costs due to the higher ammonium content of manure; labor savings due to fewer applications; potential on-farm energy savings through the construction of heat exchangers that allow energy to be recovered and used to heat residences, animal facilities, water, or dry crops. Because less nitrogen is lost through volatilization, manure with a higher nitrogen concentration may require less overall treatment with focused nutrient matching to the plants' demands. To minimize overfertilization, which can lead to leaching and run-off, the nutrient content of the manure must be considered during application. The following are some of the environmental consequences:

- ***Air***: Reduced ammonia emissions result in less particulate matter, odor, and ozone production, lowering health hazards (e.g., exacerbated asthma and respiratory disorders). Reduced greenhouse gas emissions (methane, nitrous oxide, and carbon dioxide), resulting in a lower climatic impact.

- ***Soil*:** Reduced possibility for soil acidification improved soil fertility and health.
- ***Biodiversity*:** Reduced nitrogen deposition has a positive impact on natural ecosystems.

Improving fertilization management plans: Basic fertilizer management plans could be prepared for the entire territory instead of simply large farms (> 100 ha) in nitrate susceptible zones (NVZs) and pig and poultry farms with more than 100 livestock units. Larger farms could also include more extensive calculations in their plans, rather than just assessing the fundamental nutrient balance. For example, using specific tools (e.g., soil analysis) and accounting for all relevant inputs and outputs (e.g., deposition, fertilization, crop residues), more detailed fertilization management plans could be developed, allowing the amount of nutrients applied to be optimized to the specific plot conditions (soil type, crop demand, and remaining nutrients). Nutrient losses can be significantly reduced by basing organic and inorganic fertilizer application rates on a determined nutrient balance. Determine the crop nutrient requirements for the desired yield in the given environment. Analyze the nutrient content of organic fertilizers, determine the appropriate ratio of organic to mineral fertilizers to fully meet crop needs, and consider the interval between application and nutrient uptake by crops. Reduced application of fertilizer reduces nutrition loss when done properly (including weather conditions). Another advantage is that if less additional inorganic fertilizer is applied, greenhouse gas emissions during fertilizer production are reduced. The following are some of the environmental consequences:

- ***Water*:** Reduced nutrient leaching and run-off, lessening dangers to human health and biodiversity by reducing eutrophication and enhancing surface and groundwater quality.
- ***Soil*:** Reduced acidification potential, which improves soil fertility and health.
- ***Air*:** Reduced ammonia emissions result in less particulate matter, odor, and ozone formation, lowering human health concerns (e.g., exacerbated asthma and respiratory disorders); Reduced greenhouse gas emissions (nitrous oxide and carbon dioxide), reducing climate effect.
- ***Biodiversity*:** Positive effects from reduced N deposition in natural ecosystems.

***Using appropriate application techniques*:** By reducing the surface area of manure in contact with the air, an effective manure application strategy, such as banding or injection, can limit ammonia volatilization. As a result, the potential for ammonia emissions is reduced, and the amount of nitrogen used by the crops improves. The use of nutrients can also be improved by distributing manure evenly. Nutrient run-off can be reduced when manure is incorporated soon after spreading or directly by injection. Depending on the soil type and crop, the best application technique may differ. Band spreaders (followed by integration) or injectors (open or closed slot) can be used to apply manure before sowing. Perforated hoses are dragged behind band spreaders, from which slurry is applied close to the ground. Injection systems cut slits in the soil and inject fertilizer at various depths. When spreading over grassland, a trailing shoe spreader ensures consistency and reduces emissions. There are also several bands spreading or closed slot slurry injection machines for top dressing applications. Slurries mixed with irrigation water can also be used as a distribution method, if the irrigation water does not leave the fields. Some of this equipment, such as a band spreader or an injector coupled to an umbilical system, can help optimize application time and contribute to better nutrient use. Fertilizer waste is avoided with accurate application, which decreases the need for extra synthetic N fertilizers. Odor emissions are reduced by applying close to the soil; however, injection must be timed appropriately in terms of crop needs and climate conditions to avoid an increase in nitrate leaching potential. Shallow injection may also raise the risk of nitrous oxide emissions; therefore, deep injection is preferred. Nutrient run-off is decreased by integrating or injecting manure. Due to the weight of the attachment, trailing shoe equipment may increase the risk of soil compaction. Slurry is best used in the spring, when soils are frequently moist, posing a concern of compaction. To mitigate this, umbilical slurry handling systems can be utilized, although they are costly.

Using catch/cover crops: By incorporating catch crops into the crop rotation, you may prevent nutrient leaching and run-off while also lowering your requirement for organic and mineral fertilizers. This is especially important for farms in NVZs, where the danger of nitrogen leaching must be minimized using catch/cover crops or other appropriate measures such as those outlined in the Code of Good Agricultural Practice. In order to maintain healthy soil structure and organic matter, crop rotation (including catch crops in the winter) should consider the features of each type of catch crop. To avoid further chemical application and environmental issues, catch crops should only be introduced mechanically. In some areas, incorporating catch crops is especially beneficial. On sandy soils, catch crops are more effective than on clay soils. It is vital to consider the best time for sowing and plowing under the catch crop, although there are little technical or expertise requirements. To maximize their nutrient collecting function, the catch crops should be chosen in accordance with the region and the other crops in the rotation. Most farmers already utilize legumes combined with grains, which contribute to N through biological fixation. Weeds can be controlled with catch crops (and cover crops in general). Because of their ability to fix nitrogen, legumes are a good choice for soils with low nutrient content or that are vulnerable to substantial nutrient losses. The high danger of nitrate leaching and increased nitrous oxide emissions following plowing is one disadvantage of legume-rich swards. The following are some of the environmental consequences:

- ***Soil***: Increased soil organic matter improves soil fertility and structure, reducing the danger of soil erosion and contributing to carbon sequestration and storage in the soil.
- ***Water***: Reduced danger of nutrient leaching and run-off during fallow crop times, minimizing the risk of eutrophication and enhancing surface and groundwater quality, lessening threats to human health and biodiversity.
- ***Air***: Reduced Emissions of greenhouse gases (nitrous oxide) have a reduced climatic impact.
- ***Biodiversity***: The catch/cover crops provide more wildlife and pollinator habitats, which has a positive impact.

Constructing sediment ponds to retain nutrients: Small shallow sediment basins collect soil particles and nutrients from run-off water, particularly P. This method could aid in lowering nutrient levels in surface water. A sediment pond is made up of a shallow sediment basin and a wetland that acts as a filter and is covered in typical wetland vegetation. Ponds can be created by expanding a segment of a ditch or determining the best place for a shallow basin based on the area's natural drainage. The size of the sediment pond and the applicability of the measure will be influenced by the drainage area and soil type (e.g., basins in sandy soils would not retain high structural integrity compared to other soil types). Increased soil heights along ponds may be required to slow the flow of run-off and restrict the water's course. A second wetland can be added if the region is steeply sloping to improve sedimentation efficiency. The measure does not necessitate large investments or technical expertise, but it does necessitate extensive planning prior to execution and considerable labor to construct the pond. To prevent soil erosion and nutrient losses, this strategy can be used with conservation tillage and buffer strips at the field's end. Silt ponds are low-maintenance once developed, and clearing away the built-up sediment allows farmers to recover organic matter and nutrients for reapplication to the field. This eliminates the need for additional fertilizers to be purchased and applied. Additionally, farms that use irrigation can build sediment ponds to trap and reuse irrigation run-off, increasing water efficiency. Slowing flooding and rainwater run-off improves soil and water quality by retaining nutrients in the sediment (mostly P) that are absorbed by plants. Pesticides, as well as N, P, and K linked to soil particles, can be captured. If the ponds remain wet for an extended period, there is a risk of plant death and fungal development. Sedimentation ponds emit variable levels of nitrogen and methane into the air as manmade wetlands, significantly impacting climate change. Wetlands, on the other hand, may store a lot of carbon, which helps to mitigate climate change. The following are some of the environmental consequences:

- ***Water***: Reduced eutrophication risk by storing and sedimentation run-off water containing soil particles and nutrients (particularly P, N, and K), as well as enhancing surface water quality, reducing threats to human health and biodiversity.
- ***Biodiversity***: Benefits from semi-natural habitat construction or restoration.

Increasing the height of manure heaps to increase manure compaction: The simple act of raising the height of temporary manure heaps minimizes ammonia emissions by compacting the manure under its own weight. Increasing the height of the heap to 4-6 meters would necessitate the use of special equipment, such as a telescopic loader. Although this strategy is less successful than manure cover, it is simpler and less expensive to execute.

Using an impermeable floor for manure storage: Because a large portion of the manure is stored in heaps, an impermeable floor (such as concrete) would prevent nutrient losses due to run-off and leaching. In order to expand the usage of impermeable flooring for temporary storage of solid manure and tanks for liquid manure and slurry, programs have been established.

Conservation tillage techniques: Tillage-induced erosion, both from water and wind, affects most farmers. Tillage-induced erosion causes soil loss and, as a result, nutrient loss, particularly P. The use of conservation tillage or no-till would help to prevent nutrient loss due to tillage. Conservation tillage also slows the decomposition of wastes, increasing the amount of nitrogen accessible to crops. Tillage, on the other hand, would likely have only a limited influence on nutrient losses from leaching or emissions due to the type of soil, particularly in sandy soils. This method is best suited to big or intensive farms, as well as holdings with little or no livestock. No-till may only be used in livestock-producing farms if the slurry is injected into the soil, as the slurry must be absorbed. While reduced till and no-till may yield in some labor cost reductions, they may also necessitate considerable expenditures in specialist machinery and/or result in higher herbicide and pesticide use. Weeds may also result in yield loss, resulting in additional costs.

***Incorporating crop residues*:** Incorporating agricultural leftovers or mulching enhances soil organic matter and improves soil structure and water retention capacity, lowering the danger of nutrient loss. Residues with a high carbon to nitrogen ratio, such as straw, aid in the fixation of nitrogen that is slowly released. Fertilizers combined with straw and liming improve the organic matter content of the soil. This measure has the potential to have a significant impact on regional nutrient levels. Ploughing under harvest residues is a popular way for utilizing them. Farmers employ this method to enhance their soils and because there are no other options for using straw (e.g., insufficient quantities compared to the size of the cattle and breeding systems, or since a high quantity of pig operations do not use straw for bedding).

Managing the 4 Most Important Crop Nutrients

***Controlling Nitrogen(N) Loss*:** The most vital nutrient for crop productivity is nitrogen, but it's also the most vulnerable to being lost before your crop can utilize it. Because the great majority of nitrogen is present in the atmosphere as a gas, it is unavailable to plants. Whether through commercial fertilizer manufacturing or symbiotic nitrogen fixation, gaseous nitrogen must be converted into a form that plants can use. Ammonium and nitrate are the two types of nitrogen found in soil. Because ammonium is positively charged, it is kept by soil particles and does not drain. Ammonium is transformed to nitrate as soils warm, which is negatively charged and easily leached. Nitrate loss can also be caused through denitrification, which is the conversion of nitrate to gaseous nitrogen. Controlling nitrogen loss is as simple as matching your nitrogen application to crop requirement. Crops have a low nitrogen need when they are young and tiny. Crops start a quick growth phase when the temperature rises, and the demand for nitrogen rises rapidly. By dividing the total amount of nitrogen fertilizer applied into many applications, you may ensure that the nitrogen you applied is used by the crops. Apply a tiny amount of nitrogen in your beginning fertilizer, for example, to match nitrogen supply with crop requirement. Apply numerous side-dress nitrogen treatments after that.

***Controlling Phosphorous Loss*:** The negatively charged phosphorus in the soil can bind with positively charged ions in the soil. Except in sandy soils and soils with extremely high phosphorus levels, leaching is rarely a problem. Soil erosion is a major pathway for phosphorus to get off-farm. Because phosphorus is most readily available in the soil between pH 6 and 7, adequate variable rate lime application is required to ensure that the crop receives phosphorus. Reduced tillage schemes, grassed rivers, and cover crops can all help to decrease erosion and maintain phosphorus on the farm.

***Controlling Potassium Loss*:** Potassium is a positively charged element in soil that can leach in sandy soils and locations with a lot of rainfall. Most of the potassium used in crop cultivation is administered as a dry, surface-applied substance, like potash. As a result, it's critical to get potassium into the soil where it may be used by the crop.

Controlling Sulfur Loss: Because sulfur is negatively charged in the soil, it is susceptible to leaching. Leaching susceptibility is determined by soil texture, with sandier soils being more likely to leach sulfur than clay-rich soils. Farmers with sandier soils should be aware of the risk of leaching and avoid applying large amounts of sulfur all at once. Multiple sulfur applications that are timed to coincide with crop uptake can assist ensure that sulfur is employed to produce yield rather than leaving the farm. In the absence of sulfur fertilizer, the breakdown of soil organic matter is the primary source of sulfur for crop production. Reduced tillage, the use of cover crops, and the application of compost and manure to the soil are all examples of cropping methods that retain or build soil organic matter and hence assure adequate levels of sulfur in the soil.

Dealing with Soil Acidification

Emerging acidity appears gradually, and it is often difficult to detect based on crop symptoms until it becomes a significant problem. Crop output has already been harmed for several years by the time symptoms appear. Sensitive crops like lentils and canola will be the first to show patchy development, with low vigor patches, short fat roots, poor phosphate uptake, and poor legume nodulation. The best strategy is to have a regular soil testing program that tests the top soil of each paddock every 3-5 years. Unbalanced carbon and nitrogen cycles linked with product removal, the use of acidifying fertilizers, and nitrate leaching are all contributing to continued acidification of agricultural soils. As a result, effective soil acidity management is crucial to the long-term viability of cropping operations on acid-prone soils. The following are some technical methods aimed at overcoming the limitations of acid soils:

***Techniques that avoid the problem*:** One management strategy that avoids the difficulty is the use of land for purposes other than cultivating economic crops. One such idea is urbanization; however, humanity has not been particularly successful in putting cities and towns on acid soil land. Another management method for degraded areas of little or no economic value is to conserve nonproductive land as national parks. Acid soil areas might easily rank alongside steepness, stoniness, and a lack of water as a requirement for national park designation. The designation of the acid-soil coastal wallum lands of eastern Australia as national parks, for example, might have been more prevalent if it hadn't been discovered that pine trees thrived on these soils. Policymakers have been more active in reserving land for forestry, and because many tree species thrive on acid soils, this is a well-defined technique for soil acidity management. A strategy like this is applied at three levels: at the government level in defining policy and as a forest manager, at the business level as a forest manager, and at the individual farmer level. In some areas with severe deforestation, the price of timber products has risen to the point where smallholder farmers can afford to grow trees for wood rather than annual crops. Farmers use spatial variability in soil pH if population pressure is low and the amount of land available per farmer is sufficient. Cropping patterns, for example, are ordered according to soil acidity in sections of East and Southern Africa's gently sloping plateau areas. Crops adapted to acid soils are planted on the upper parts of the landscape, whereas more nutrient-demanding and acid-soil sensitive crops are produced on the lower slopes, where the soils are less acid and have a greater nutrient status.

***Traditional techniques in developing countries*:** Traditional shifting cultivation techniques rely on the role of fire to quickly reintroduce nutrients from natural plants into the soil. Ashes serve as a fertilizer as well as a lime amendment. All the 'slash and burn' strategies, on the other hand, only provide a brief time of adequate fertility for crop production, usually 1-3 years. As a result, the 'slash and burn' systems are still low-intensity 'shifting systems.' To restore natural fertility to its original levels, long fallows are required. The majority of these 'slash and burn' techniques are only acceptable in areas where land is not a constraint. Because the Chitamene system's fertility-import principle necessitates the clearing of a woodland area ten times the area cropped, it is known for its environmental impact. Adaptations of fire-based practices, on the other hand, have happened in various parts of Africa where land is scarce. The lesser-known approach of' ecobuyage,' which is based on slow combustion under decreased oxygen availability, is an adaptation to a more settled agriculture because the fertility increase and toxicity alleviation appear to continue longer.

***Techniques that use better-adapted plants*:** The goal of these procedures is to alter the plant or change it from one type to another. If a soil becomes too acidic for a typical crop to function well, other, better-adapted crops should be considered. Farmers are familiar with these crops and have practiced the method of developing adapted crops for generations. However, due to the possibility that the adaptation is limited to a single component of the soil acidity complex - AI, H or Mn, or low P - the adapted plant may perform well on one acid soil but not on another, some caution is advised. On the other side, there are cultivars that are both low pH and AI tolerant. Also, within a species, there may be enough genetic variability to classify specific kinds as acid-tolerant or acid-adapted. Soybean and sugarcane are thought to be acid-tolerant in some regions, but thrive in acidic soils elsewhere. There is a quest for adaptable cultivars where there is scientific support. Some of this activity may have transpired without anyone's knowledge, as it appears to have in the Australian sugar business, where high sugarcane yields are attained on acid soils. Sugarcane breeders have unwittingly chosen for acid tolerance or avoidance by locating their screening trials on acid soil areas. One issue with this strategy is that development in acid soils is constrained by a variety of factors. Tolerance or avoidance are two mechanisms that can be used to deal with these limits.

- Low internal mineral nutrition demand and toleration of hazardous substances are examples of tolerance mechanisms.
- Roots' ability to modify the rhizosphere, eliminate hazardous substances from absorption locations, and boost their ability to take up limiting nutrients are all examples of avoidance mechanisms.

None of these features are present in every species or cultivar. Tolerance to high Al concentrations does not imply adaptability to other harmful elements or the ability to cope with low critical element concentrations. As a result, a plant or cultivar chosen for its capacity to thrive in one acid soil may struggle in another. Either numerous adapted types must be discovered and screened for each situation, or the nature of the adaptation or the soil must be understood in order to select the appropriate plant. Farmers have long practiced selecting adaptable species. This can be demonstrated in a variety of settings, such as the usage of perennials, such as rubber, oil palm, coffee, and other tree crops, which have had some significant triumphs in the tropics. Pineapple and cassava are among the best examples of short-term crops. Such success comes with a cost, as overproduction leads to low commodity prices, as it has in the case of coffee and rubber, and may soon be the case with pineapple. Growing the best-adapted crop may be discouraged by government pressure in some nations, as it is with poppy in Southeast Asia, where finding alternative, well-adapted income crops has proved difficult. Lupin's adaptation to acidic Australian wheat soils, on the other hand, means that most of the area that was formerly mono-cropped to wheat now has a legume crop in the rotation.

Table 18: Farm crops that may grow well in acidic soils

Type	Tropical	Temperate
Forage grasses	*Brachiaria decumbens* *Panicum maximum* *Andropogon gayanus*	*Agrostis tenuis* *Cynodon dactylon* *Paspalum notatum*
Forage legumes	*Centrosema pubescens* *Pueraria phaseoloides* *Stylosanthes capitata*	*Trifolium subterraneum*
Food crops	Coffee Pineapple Rice	Potato Blueberry Cranberry
Food legumes	Peanut	Lupin

Source: Myers, & De Pauw, (1995).

Techniques that treat the soil problem: The goal of these strategies is to change the soil. This was formerly accomplished in the tropics by spelling the land, as evidenced in traditional slash and burn agriculture. The land reverts to forest after 1-3 years of cropping, allowing the soil to replenish its organic matter, allowing successful agriculture to resume after a few years. This system was thought to be an environmentally friendly low input-low output technology. With growing land pressure and shorter rotations, this method has become less practicable, especially on steep terrain, where it also has downstream consequences. At least one Southeast Asian country expects a significant decline in hydropower capacity due to dam siltation caused by acidic upland soil erosion. Farmers and scientists are now working to keep production continuing without reverting to a forest or grassland fallow. The causes of low output in these circumstances must be addressed by importing or cultivating supplies that can rectify toxicities and nutrient deficits. The inputs are either those that raise the soil pH, such as lime, or those that reduce the concentration of toxic ions in the soil solution without raising the pH significantly, such as organic materials that decompose to produce certain organic acids, or those that generally increase soil organic matter while removing Al from the soil solution, such as organic materials that decompose to produce certain organic acids, or those that overcome nutrient deficiencies, such as organic materials that decompose to produce certain organic acids.

***Improved traditional techniques*:** Composting has only been widely employed as a way of fertility maintenance in traditional Chinese communities. In the tropics, ley systems are uncommon, in which multiple years of arable cropping are followed by one or more years of grazed grass and legumes or restorative fallow. Even though there are potential green manure species that are well adapted to acid soils, small farmers in the tropics rarely use them. Farmers are often reluctant to cultivating a crop that takes up land for the entire or a significant portion of the growing season, provides no direct return, and necessitates labor for planting and digging-in at a time when food crops require labor.

***Modern techniques - liming, fertilizers, and organic matter*:** These have mostly relied on the use of liming materials to minimize Al and perhaps Mn toxicity. Limestone is the most prevalent material, but marl, slag, and other materials are also utilized. They contain different amounts of magnesium and silica, with the silica added by unscrupulous operators in some cases. Limestones range in composition from dolomite (12-13 percent Mg) to high Ca limestone (less than 1% Mg). In Mg-deficient settings, the presence of Mg becomes critical. The use of fertilizers is the second pillar of modern acid soil management. Fertilizer rates should ideally replenish nutrients lost by crops, but due to P fixation and low K retention, rates may not always result in maximum yields. Rock phosphates can have a liming impact in acid soils because they deliver Ca, increase the pH, and possibly lower the exchangeable Ai at a low cost. Rock phosphate has a wide range of reactivity, which can be improved by fine grinding. Organic matter control is the third pillar of current acid soil management. Recognizing that soil loss of organic matter was a contributing component in acidification, efforts were made to control acidity by attempting to reverse organic matter loss. However, in some systems, organic matter accumulation is accompanied by nitrogen fixation, and the enhanced nitrification rates that result cause acidification. Raising soil organic matter levels with organic inputs has proven highly difficult and impractical in farmed systems, particularly in tropical locations. Organic matter inputs, on the other hand, may still have a role in decreasing the impacts of soil acidity. Green manures, composts, and other organic wastes have been shown to minimize soil Al toxicity while also increasing crop yields.

Dealing with Soil Salinization

When evaluating salinity management solutions, it's important to distinguish between applied water salinity and soil salinity. Crop salinity tolerances are measured as a combination of average applied water salinity and average root zone soil salinity over the course of a season. Salts are carried by irrigation water, and when it is applied to fields, salts are added to the soil. Salts build up in the soil at larger concentrations than they did in the water that was applied. Salts may also build up unevenly in the soil. For all these reasons, testing water and soil salinity on a regular basis is critical to understanding baseline conditions and changes over time. An electrical current is sent through a soil solution taken from a saturated soil sample to determine salinity. Electrical conductivity refers to a solution's capacity to convey a current (EC). The deci-Siemens per meter (dS/m) unit of measurement for EC is the numerical counterpart of the former millimhos per centimeter unit of measurement (Table 2). The dS/m rating decreases as the salt content of the soil decreases, and the effect on plant growth decreases. Soil samples taken for soil fertility objectives are suitable for examining for salinity trends, but they may not provide the same information as salinity sampling done specifically for diagnosis. For more information on analyzing saline soil conditions, see table 19, Diagnosing Saline and Sodic Soil Problems. Electromagnetic conductivity meters can also be used to measure soil salinity in the field. This device works by inducing a magnetic field in the soil with the help of a transmitting coil. Induced currents in the soil are detected using a receiving coil. This equipment is usually trailed or dragged behind an ATV and takes readings as it drives through the field in a pattern. A map of salinity levels can be created using GPS technology to display relative salinity levels within a field. With this knowledge, salinity control measures like as site selection, variety selection, soil amendments, and leaching can be implemented.

Table 19: Relative salt tolerance of a range of vegetables

Crop	Salinity threshold EC_{se} (dS/m)	Soil salinity EC_{se} at 90% yield (dS/m)
Pea	1.0	–
Bean	1.0	1.5
Carrot	1.0	1.7
Onion	1.2	1.8
Lettuce	1.3	2.1
Pepper	1.5	2.2
Corn (grain sweet)	1.7	2.5
Potato	1.7	2.5
Tomato	2.3	2.8
Cabbage	1.8	2.8
Spinach	2.0	–
Watermelon	–	2.0
Cantaloupe (rockmelon)	2.2	–
Cucumber	2.5	3.3
Broad bean	2.3	–
Celery	1.8	3.1
Broccoli	2.8	3.9
Squash	4.7	5.8
Zucchini	4.7	5.8
Garden beet 4.0 5.1		

Source: Shaw et al., (1987).

Leaching: The greatest technique for preventing deleterious salt accumulation in the soil profile is to leach excess salts and maintain a positive salt balance. This is accomplished by providing enough water to allow salts to seep below the root zone without avoiding groundwater reserves.

Drainage: Good internal and exterior drainage is a need for employing leaching as a management strategy. Tillage and soil amendments are frequently employed to address poor internal soil drainage caused by surface crusting, hardpans, and sodic conditions. In these circumstances, deep ripping should be done on a regular basis. When sodic conditions occur, an active soil amendment program, such as the use of gypsum, is required.

Surface drainage is critical, especially when using furrow irrigation. Growers should be mindful of crop water requirements to minimize over-irrigation, and laser leveling is a usual advice. Internal drainage is being improved by the increased use of tile drains in horticultural plantings (and mole drains in vegetable farming). Before planting new regions, a thorough soil survey is highly recommended.

The irrigation method: The irrigation method used and the amount of water used have a big impact on salt accumulation and distribution. Salts are generally moved below the root zone by flood irrigation and a sufficient leaching fraction. A correctly controlled sprinkler irrigation system can produce similar outcomes. Soluble salts in the soil migrate along the wetting front in furrow and pressured irrigation, concentrating at its termination or at its convergence with another wetting front. Water travels away from the emitter in drip-irrigated plots, concentrating salts where the water evaporates. Water moves from the furrow into the bed by capillary flow in furrow-irrigated plots. Salts collect in the center of the intervening bed when adjacent furrows are irrigated. In furrow-irrigated row crops, manipulating bed form and planting arrangement are common tactics for avoiding salt damage. Plants tolerate higher amounts of salinity with drip irrigation than with furrow irrigation because the soil moisture conditions are more consistent.

Fertilizer management: Many fertilizers have large quantities of soluble salts. As a result, nutrition source, rate, timing, and positioning are all key factors to consider when growing horticultural crops. Most commercial fertilizer formulations have salt indices available. The salt index of KCl, for example, is 205 times that of K_2SO_4. In general, fertilizers with high salt indices should not be applied in bands near seedlings. Other salt-containing substances added to the soil, such as gypsum and manures, should also be considered. Using gypsum to prevent sodium buildup on the soil's exchange complex, maintain soil structure, and improve water infiltration is a good management strategy. Apply gypsum well before seeding salt-sensitive crops like lettuce so that soluble salts generated during dissolution do not negatively affect yield.

Water treatment and soil amendments: Many salty and sodic soil problems can be solved using soil amendments and water treatments, which are both practical and cost-effective. Soil additives are utilized for both initial reclamation and long-term soil quality maintenance. Water applications are designed to change the chemistry of irrigation water so that soil quality does not deteriorate further. The rates at which soil amendments are applied are often high and are mostly determined by economics. The rates of modifications in water treatment are often substantially lower and are almost always based on solubilities. For years, gypsum and elemental S have been utilized as amendments. Gypsum is typically employed on Na-affected soils as a source of Ca^{++} ions to replace Na^{+} ions, which disperse soil particles and limit water infiltration. The displaced Na^{+} ions are easily leached out of the soil profile. Gypsum is a non-acidic salt that does not reduce pH directly. However, by lowering the hydrolysis events associated with Na^{+} ions on the exchange complex, it can indirectly lower the pH of sodic soils.

***Incorporation of organic matter*:** The repair process is aided by maintaining good infiltration and permeability levels. This can be accomplished by adding organic matter to the soil, leaving harvest leftovers on the surface, increasing soil surface cover/mulching, and using minimal tillage techniques. It is generally understood that maintaining an increase in organic matter in the soil has a direct impact on the capacity of the soil to hold water, as well as an indirect impact on water infiltration and circulation within the soil.

***Physical operations*:** Subsoiling (deep plowing and deep ripping) can help with subsurface sodic layer repair. This operation must be meticulously managed to avoid causing other issues, such as rhizosphere water saturation during crop growth. Although it is an expensive process, mixing sand with thick textured soil can improve soil physical qualities and boost drainage capacity and infiltration rate. As a result, the soil reclamation process is improved, and salt leaching is improved. To avoid having a negative impact on the soil structure and fertility, this must be properly planned. For small regions, physical removal of the surface salt crust may be sufficient. The salts that develop at the soil surface due to capillary rise and subsequent evaporation can be manually or mechanically removed, albeit this is a cost-ineffective method that may have a short-term effect.

***Plant selection*:** Some crops have a strong tolerance for salt, with productivity drops of only 10% at ECE levels of 8 dSm-1 or higher. Olive and rapeseed, for example, are examples of this. New cultivars are being created to adapt well with high salt levels. In some very high salinity environments, halophyte plantation is also becoming an agronomical niche. A summary of crops and types that can be utilized as a saline soil adaptation is presented in the Mini-paper Examples of salt resistant crops as an alternative for farmers.

***Crop rotation*:** The selection of an appropriate crop rotation can be used as a response to soil salinization. During the wet season, the less tolerant species can be used, whereas during the dry season, more tolerant crops can be cultivated. A set-aside period would occur at the start of the rainy season, when the rain washes away the salts and reduces salinity to levels that the crops can tolerate. Irrigation could also be employed on alternating years in more severe circumstances. Rain may not always be enough to wash away the salts supplied during the irrigation season during dry winters. To minimize additional salinization, the salinity of the soil in the root zone must be evaluated on a regular basis.

***Microbial management*:** The microbial communities of naturally salt-affected soils show promise as a source of beneficial microorganisms and a gene bank for future biotechnological uses. From a genetic standpoint, these species have an under- or over-expression of specific genes and metabolites that enable them to deal with osmotic stress. Different types of saline-related microorganisms from various taxa of bacteria (*Baciullus, halobacillus*, cyanobacterial filaments), archaea, and fungi can be used to make bio-inoculant to promote and improve salt tolerance in plants. They can also be employed in the restoration or prevention of salinized soils. Several researches revealed the good impacts and efficacy of microbial salt stress amelioration, which involved inoculating crop seeds, soil, and farmyard manure applied to soil with specific bacteria strains.

***Synthetic conditioners for maintaining the soil structure*:** Due to their enormous size, which allows them to be absorbed concurrently by numerous soil particles, compounds like polyacrylamides (PAM) and polysaccharides operate as cementing agents binding soil particles and boosting soil structural stability. Low doses are required, for example, to prevent soil crusting and erosion in furrow and spray irrigation, roughly 10-20 kgha-1 of PAM is sufficient. There are various commercial brands on the market, but their economic cost makes them only appealing in the event of crops with extremely high added value.

***Phyto and bioremediation of sodic and saline-sodic soils*:** By making Ca available to replace Na in the soil's exchange complex, phytoremediation can be employed for Na removal in a similar way to chemical amelioration. Other processes, such as halophytes' salt bioaccumulation in aboveground biomass, have been demonstrated to be ineffective in lowering soil salinity. As a result, the method of salt removal via phytoremediation necessitates a source of Ca, which is commonly calcite found in soils. The plants' involvement in this process is to boost CO2 levels in the root zone, which aids calcite dissolution. Bacterial activity can help elevate CO2 levels in the root zone even more. In circumstances with low to medium sodicity, phytoremediation can be applied. Studies have demonstrated that phytoremediation can produce equivalent or better results than chemical remediation, but it necessitates more detailed and complicated planning during the remediation process (crop rotation, irrigation timings, etc.). Farmers may find phytoremediation to be a more cost-effective mitigating option. It's also better for the environment because it doesn't require synthetic products and improves carbon sequestration. As a result of root exudates and calcite dissolution, studies have demonstrated that it can also improve total nutrient availability to plants. When opposed to chemical treatment, Na removal happens more consistently and in-depth because it occurs along the root depth; for example, alfalfa roots can reach 1.2 m deep.

Chapter 4

Fertilizers and Soil Amendments

CONTENTS

INTRODUCTION

Fertilizers and soil amendments are a variety of compounds that are given to soils to help plants grow more quickly. They can be natural, like bone meal, or synthetic, like 10-10-10 fertilizer. Some must be purchased, while others can be taken for free from your surroundings. Many are dual-purpose, acting as fertilizers as well as soil amendments. Anything placed into topsoil to support healthy plant growth is referred to as a soil amendment. They serve a variety of purposes. They could, for example, alter the pH of the soil or provide nutrients. Fertilizers are valued primarily for their ability to provide nutrients. Plants employ these minerals to build proteins and carbohydrates, which are essential for plant growth. Soil conditioners, such as composted horse manure, are a type of soil additive that improves soil structure by binding soil particles into bigger aggregates. This increases pore space, allowing for better air exchange, water circulation, and root growth.

FERTILIZERS

Introduction

Mineral and organic fertilizers are two types of fertilizers. Mineral fertilizers include mostly nitrogen (N), phosphate, and potash, and are also known as chemical fertilizers since they are manufactured by the chemical fertilizer industry. Although natural fertilizer deposits exist, practically all nitrogen fertilizers are made by the Haber–Bosch process from nitrogen gas (N2). Phosphate fertilizers, on the other hand, are made from the acid attack of natural rocks, whereas potash fertilizers are made from soluble potassium salt reserves (e.g., potash ores and brines). Most organic fertilizers come from animal manures — either the feces and urine of cattle animals or the feces of poultry, either alone or in combination with bedding material (e.g., farmyard manure, chicken litter). Plant residues, municipal effluents and biosolids, and organic by-products are among the other sources. Organic manures and leguminous N fixation have been the primary sources of nutrients for farmers for thousands of years, contributing greatly to agricultural productivity and soil quality. The usage of mineral fertilizers rose exponentially following the discovery of the Haber– Bosch process at the turn of the twentieth century, a rapidly rising human population, and dramatic improvements in agricultural science such as the Green Revolution. In 1950, roughly 14 million nutrient tons of fertilizer were used, compared to 142 million nutrient tons in 2002. World cereal production increased from

740 million tons to 2085 million tons within the same time period. These few numbers clearly show that mineral fertilizers have played a critical role in agricultural production systems' success. The continued exponential growth of the human population, as well as the growing demand for biofuels, indicate that fertilizer demand will continue to rise. Despite the apparent success of present agricultural production systems, excessive fertilizer use has resulted in serious environmental issues and an increase in health concerns. The Green Revolution's 'evil side' was its single-minded goal of expanding food production. Inadvertent environmental concerns resulting from the overapplication of mineral fertilizers were initially overlooked. Overapplication was once regarded to be an insurance policy against yield-limiting nutrient losses, and it still is in many regions. The relatively low cost of fertilizers, as well as the lack of environmental protection legislation, encouraged this conduct. Increasingly affluent civilizations have also modified their eating habits, with meat consumption skyrocketing. Over the last 30 years, the animal agriculture industry has seen substantial changes as a result of this demand. Bigger and larger concentrations of animals have resulted from economies of scale. Concentrated animal feeding operations (CAFOs) or, more euphemistically, "factory farms" are terms used to describe these facilities. Environmental contamination issues related with the disposal of the generated manures have been exacerbated by the rising concentration of animals. Land application is increasingly being used to capture the nutritious value of manures, as well as to dispose of them. Nonetheless, manure overapplication is a continuous issue with serious environmental effects. Overall, there are two main issues that cause environmental and human health risks when fertilizer is overapplied:

- the disruption of the natural N and phosphorus nutrient cycles because of the massive infusion of N and phosphorus (P) fertilizers; and
- the inadvertent release of heavy metals and radionuclides from mineral fertilizers and pathogens, veterinary pharmaceuticals, and endocrine disrupters from organic fertilizers.

Soil Fertility Improvement and Plant Nutrient Needs

Fertile and productive soils are essential components of stable communities because they allow plants to grow that are needed for food, fiber, animal feed and forage, industrial products, energy, and a beautiful environment. Soil fertility combines the fundamental principles of soil biology, soil chemistry, and soil physics to establish the strategies required to manage nutrients profitably and sustainably. The ability of soils to supply plant nutritional requirements varies greatly; most have only moderate natural soil fertility. In order to meet production goals, more nutrients are frequently required than the soil can provide. High crop yields deplete soil nutrient resources, which must eventually be compensated for by higher nutrient input to sustain the fertile soils that modern societies require. Soil fertility regulates the supply of nutrients inherently available in soils or applied as manures and fertilizers to plants, and it is a hallmark of high-intensity agriculture's reliance on mineral fertilizers to restore soil fertility. In the broader context of soil productivity, soil fertility regulates the supply of nutrients inherently available in soils or applied as manures and fertilizers to plants. Even without added fertilizer, soils with high natural fertility can produce large crop yields, but with an additional supply of the key nutrients, they can produce even higher yields. Soil fertility is critical to successful farming and should not be overlooked.

Plants include nearly all of the natural elements (92), but 17 of them have been identified as important nutrients for plant growth. Mineral fertilizers or the soil must offer these, as well as plant and animal wastes and/or other organic sources. To prove that an element is important, it must be demonstrated that a plant cannot complete its life cycle without it, and that no other element can replace the test element. Carbon (C), hydrogen (H), and oxygen (O) are the three most often used elements, and they are all provided by air and water. The remaining 14 nutrients are mineral components that the plant roots get from the soil. Plants require relatively significant amounts of the three macronutrients. Nitrogen is a non-reactive gas that makes up 78 percent of the Earth's atmosphere. To be used by plants, it must be transformed to reactive chemical forms (ammonium and nitrate). Microorganisms in the soil, symbiotic bacteria on plants, and chemical reactions all contribute to this conversion. Phosphorus (P) is abundant in soil minerals and organic matter, but it must be transformed to inorganic phosphate ions ($H_2PO_4^-$ or HPO_4^{2-}) before plants can use it. Potassium (K) is abundant in soil minerals and adsorbed to soil particles and organic matter in the ionic form K+. As a K+ ion, it enters the plant roots as a companion to negatively charged ions through osmosis through cell walls. In plants, potassium does not produce any chemical compounds, but it is essential for the transfer of water and other ions through cell membranes. The three secondary macronutrients, sulphur (S), calcium (Ca), and magnesium (Mg), are just as important for plant growth as the macronutrients, but in smaller levels. Sulphur is found in organic materials in the soil, as well as various clay minerals. Plants absorb sulphur as the sulphate ion (SO_4^{2-}). Calcium and magnesium are readily available in the soil and are readily absorbed by plant roots as cations. Mg, being a core component of the chlorophyll molecule, plays a major role in photosynthesis and is an important structural component of cell walls and plant tissues. The micronutrients are the eight essential minerals that plants require in small amounts and include iron (Fe), zinc (Zn), copper (Cu), manganese (Mn), molybdenum (Mo), chlorine (Cl), boron (B), and nickel (Ni) (Ni). Cobalt (Co) and silicon (Si) are two more nutrients that are required by some plant species but not by others.

Table 20: Essential and beneficial mineral nutrients for plants

Category	Nutrient	Intake Form	Source
Macronutrient	Nitrogen (N)	nitrate, NO^-_3; ammonium, NH^+_4	organic matter
	Phosphorus (P)	phosphate, HPO_4^{2-}, H_2PO_4	organic matter, minerals
	Potassium (K)	potassium ion, K^+	minerals
	Calcium (Ca)	calcium ion, Ca_2^+	minerals
	Magnesium (Mg)	magnesium ion, Mg_2^+	minerals
	Sulphur (S)	sulphate, So_4^{2-}	organic matter, minerals
Micronutrient	Chlorine (Cl)	chloride, Cl-	minerals, rainfall
	Iron (Fe)	ferrous iron, Fe_2^+	minerals
	Boron(B)	boric acid, H_3BO_3	organic matter
	Manganese (Mn)	manganese ion, Mn_2^+	minerals
	Zinc (Zn)	zinc ion, Zn_2^+	minerals
	Copper (Cu)	cupric ion, Cu_2^+	organic matter, minerals
	Molybdenum (Mo)	molybdate, MoO_4^{2-}	organic matter, minerals
	Nickel (Ni)	nickel ion, Ni_2^+	minerals

Source: Johnston, & Bruulsema, (2014).

To obtain these elements, many sources of nutrients can be used, with fertilizers being the most used technique to maintain soil fertility.

Inorganic/Mineral Fertilizers

A variety of mineral fertilizers have been created to enhance nutrients already present in the soil and to meet the high demands of crops. Except for some organic substances like urea, which are easily transformed into salts, these are mostly mineral salts. The traditional division of fertilizers into single and multi-nutrient fertilizers usually only pertains to the three primary nutrients. Many ostensibly single-nutrient fertilizers really include multiple nutrients; for example, ammonium sulphate contains both N and S. Fertilizer grade is a system for categorizing fertilizer ingredients based on their composition of the three major nutrients. The nutritional content, or grade, can refer to the total or accessible nutrient content, and can be stated in oxide form (P_2O_5, K_2O) or in elemental form for specific nutrients (N, P, K). The following are the several types of mineral fertilizers:

- ***Method of production***: Natural (as found in nature or with minimal processing); synthetic (as found in nature or with minimal processing) (manufactured by industrial processes).
- ***Number of nutrients***: multi-nutrient (multiple nutrient) or compound fertilizers, with 2, 3 or more nutrients: single-nutrient or straight fertilizers (whether for major, secondary, or micro nutrients); multi-nutrient (multiple nutrient) or compound fertilizers (whether for major, secondary, or micro nutrients):
- ***Type of combination***: Complex fertilizers, in which two or more nutrients are chemically combined; mixed fertilizers, which are a physical mixture of two or more single-nutrient or multi-nutrient fertilizers (for granular products, this may comprise a blend of separate granules of the individual ingredients, or granules each containing these ingredients); (e.g., nitro-phosphate, ammonium phosphates).
- ***Physical condition***: Solids of various sizes (crystalline, powdered, prilled, or granular); liquids (solutions and suspensions); gaseous (liquid under pressure, e.g., ammonia).
- ***Mode of action:*** slow-acting (water-soluble and instantly available); quick-acting (water-soluble and immediately available) (transformation into soluble form required)

Straight fertilizers: Straight fertilizers contain only one primary plant nutrient, such as nitrogen, phosphorus, or potassium. Urea, ammonium sulphate, potassium chloride, and potassium sulphate are just a few examples.

***Complex fertilizers*:** Two or three primary plant nutrients are present in complex fertilizers, with two of the primary nutrients in chemical combination. Diammonium phosphate, nitro-phosphates, and ammonium phosphate are examples of these fertilizers that are often produced in granular form.

***Mixed fertilizers*:** These are fertilizer blends that are made up of straight fertilizers in a physical form. They have two or three major plant nutrients in them. The materials are fully mixed, either mechanically or manually, to create mixed fertilizers.

***Solid fertilizers*:** Solid fertilizers are in several forms viz.

- Powder (single superphosphate),
- Crystals (ammonium sulphate),
- Prills (urea, diammonium phosphate, superphosphate),
- Granules (Holland granules),
- Super-granules (urea super-granules) and
- Briquettes (urea briquettes).

***Slow- and controlled-release fertilizers*:** CRF stands for controlled-release fertilizer, which is a granular fertilizer that slowly distributes nutrients into the soil (i.e., with a controlled release period). Controlled-release fertilizer, also known as controlled-availability fertilizer, delayed-release fertilizer, metered-release fertilizer, or slow-acting fertilizer, is a type of fertilizer that releases nutrients gradually. CRF is a term that usually refers to nitrogen-based fertilizers. Slow- and controlled-release fertilizers account for approximately 0.15 percent of the fertilizer market (562,000 tons).

Nitrogen Fertilizers

Nitrogen fertilizers come in a range of formulations, each with its own set of qualities and applications in crop production. These all start with anhydrous ammonia, which is produced by the Haber-Bosch process from air and natural gas. Distinct nitrogen fertilizers are rated based on their overall nitrogen content, different N forms (which impact the rate of action), and any negative effects. Regardless of the fertilizer's formulation, most of it is transformed in the soil to nitrate and ammonium, the most plant-available forms of nitrogen. Nitrate N in soil solution is readily available and works quickly, but it is most susceptible to losses due to leaching and/or denitrification. Plants mostly absorb nitrogen in the form of nitrate. Because ammonium-N is initially adsorbed on soil particles and then gradually released and nitrified, it has a slightly slower effect, although being fully available. Because N in the ammonium form bonded to soil particles is far less vulnerable to leaching and other losses, this can help with N use efficiency. Some plants can immediately absorb ammonium, whereas others must first convert it to nitrate. An application of 50-100 kg/ha (20-40 lb/A) N would nitrify in about two weeks at a temperature of 20-25° C. By adding nitrification inhibitors to the fertilizer, nitrification can be delayed over several weeks. This could be effective for limiting unwanted nitrate accumulation in vegetable crops or lowering leaching losses. To assist farmers, handle fertilizer N more efficiently, a variety of formulations, coatings, and additions are available. Stabilizers, inhibitors, slow-release, and controlled-release medicines are some of the most common types. These goods are characterized as follows by the Association of American Plant Food Control Officials (Trenkel, 2010):

- ***Slow- or controlled-release fertilizer*:** A fertilizer containing a plant nutrient in a form that delays or extends its availability to the plant substantially longer than a standard "rapidly available nutrient fertilizer" such as ammonium nitrate or urea, ammonium phosphate, or potassium chloride. A multitude of causes can cause such a delay in initial availability or an extended period of continuing availability. Controlled water solubility of the material can be achieved through semipermeable coatings, occlusion, protein materials, or other chemical forms, or through gradual hydrolysis of water-soluble low molecular weight compounds, among other methods.
- ***Stabilized N fertilizer*:** A nitrogen stabilizer has been added to a fertilizer. A nitrogen stabilizer is a material added to a fertilizer that extends the time the fertilizer's nitrogen component (urea-N or ammoniacal-N) remains in the soil.
- ***Nitrification inhibitor*:** A chemical that prevents ammoniacal-N from being converted to nitrate-N in the body.
- ***Urease inhibitor*:** A chemical that prevents the enzyme urease from hydrolyzing urea.

Anhydrous ammonia is then reformulated into a variety of additional nitrogen fertilizer sources, giving farmers a variety of alternatives for managing nitrogen to best meet their crop needs while also satisfying logistical requirements. The following are some of the more prevalent N formulations.

Anhydrous ammonia (NH_3): The most concentrated commercial fertilizer N source is anhydrous ammonia (82 percent N). Because natural gas (methane) is the most frequent source of energy for producing ammonia, ammonia production plants are typically placed near natural gas supplies. Pipelines, trucks, trains, and ships transport ammonia as a liquid under pressure and/or refrigeration around the world to keep it below its boiling point (-33°C, -27°F). Ammonia is commonly injected into the soil at a depth of 10 to 20 cm (4 to 8 in) as a pressurized liquid that vaporizes quickly and combines with the soil water to form ammonium (NH_4^+). The negatively charged cation exchange site on clay minerals and organic materials in the soil attracts this ion. Mixing ammonia with water yields aqua ammonia (20 to 24 percent nitrogen). As an alternative method of application, this form can be added to irrigation water.

Ammonium sulphate [$(NH_4)_2SO_4$] (21% N): This is one of the oldest manmade N fertilizers and is created as an industrial byproduct. It comes from the production of steel, nylon, and other sulphuric acid-based products. It's frequently employed as a carrier for herbicide application, which improves efficacy. It also includes 24% S, making it a good option when S is required.

Urea (46% N): Urea is the world's most extensively used solid nitrogen fertilizer. The controlled reaction of ammonia gas (NH_3) and carbon dioxide (CO_2) with enhanced temperature and pressure is used to make urea fertilizer. With sophisticated granulation machinery, liquid urea is shaped into spheres or hardened into a solid prill while falling from a tower. Two urea molecules may accidentally join during the synthesis of urea to generate biuret, which can be harmful when sprayed over plant foliage. Due to highly controlled manufacturing processes, most commercial urea fertilizer includes only trace levels of biuret. Urea is a good nutrition source for plants to meet their nitrogen requirements. Surface-applied urea gets into the soil with rainfall or irrigation because it dissolves easily in water. Urea travels readily in the soil with soil water until it is hydrolyzed to create NH_4^+.

Nitro-phosphate (variable grades): This is manufactured by using nitric acid instead of sulphuric acid to treat rock phosphate. It has the advantage of not generating the calcium sulphate (gypsum) byproduct, which can be problematic to dispose of. Calcium nitrate and calcium ammonium nitrate are two more byproducts produced throughout the procedure. Nitro-phosphates can be combined with other nutrients to create homogenous fertilizer pellets that are multi-nutrient.

Ammonium nitrate (NH_4NO_3): Ammonium nitrate was first manufactured as a munition in the 1940s. It has a nitrogen content of 33 to 34 percent. By reacting ammonia gas with nitric acid, ammonium nitrate is generated as a concentrated solution. The solution (95 to 99 percent ammonium nitrate) is sprayed onto small grains in a revolving drum and solidifies into prills, which can be used as fertilizer or manufactured into granular ammonium nitrate. Because half of the nitrogen is in the form of ammonium, it can be taken up directly by roots or converted to nitrate by bacteria over time, resulting in a delayed release of nitrogen. The other half of the nitrogen is in the form of nitrate, which is readily available to plants. Because of its high solubility, it's ideal for fertigation and foliar application.

Urea ammonium nitrate (UAN) (28% N): UAN is extensively used as a broadcast application of liquid fertilizer N, as a carrier for herbicides, and as a side-dress application for row crops like maize.

Calcium cyanamide: Calcium cyanamide possesses herbicidal and fungicidal characteristics as a result of intermediate breakdown products, in addition to its fertilizer value.

When different types of nitrogen are delivered to soil, crop yield responses are nearly identical. Under temperature and soil moisture conditions, the efficiency of some products may be lowered due to nitrate leaching losses or ammonia volatilization. Urea or UAN solutions applied to the surface are particularly vulnerable to such losses. Most N fertilizers are readily available and prone to loss before the N is absorbed by the crop. Slow-release or controlled-release enhancement products, on the other hand, can assist lessen those losses. Because nitrogen fertilizers tend to raise soil acidity, pH may need to be corrected using liming agents as N fertilizer use increases.

Nitrogen Use Efficiency

Depending on the type of crop, climate, soil, and management practices, crops can use up to 50% of the nitrogen provided through fertilizers. Wheat grown in temperate areas can have a yield of 50-60%, while lowland rice grown in coarse textured soils can have a yield of roughly 30%. The energy required to manufacture fertilizer N for application per unit area is roughly one-third of the total energy required to grow the crop. As a result, more efficient use of nitrogen fertilizers results in a net energy savings. Excess nitrogen that isn't used by the crop is affected by biological, chemical, and physical processes. Weather, soil conditions, and other factors influence their proportional impact on crop N supply. Fertilizer best management practices (FBMPs) for the application of plant nutrients aim to improve nutrient efficiency while reducing negative environmental consequences. In any given year, the root system of most arable crops only explores 20-25 percent of the available soil volume. As a result, plant nutrient use is influenced not only by growth stage and nutrient requirement, but also by the rate at which plant nutrients are delivered to the root via mass flow and diffusion in the soil solution.

Split application: Multiple applications of nitrogen fertilizers during the growth season can aid in improving N use efficiency and reducing losses. Applying N fertilizer as soon to the time of crop uptake as possible is a solid management method for increasing efficiency. Similarly, site-specific fertilizer management ensures high fertilizer N usage efficiency by applying fertilizer N after taking into consideration the soil's N supplying capacity. Leaching and denitrification are likely to remove any leftover mineral N in the soil after harvest. Cover crops and agricultural residue management can help to keep nitrogen in organic compounds in the soil and reduce leaching and denitrification losses. There are several technologies available to help improve the efficiency of nitrogen fertilizers. Chemical additives, biological inhibitors, and coatings that physically limit N activity in the soil are examples of these. Microbial activity is required for several critical conversion activities in the soil. These give a point of control for microbial activity by chemical or physical parameters. Here are a few examples:

- ***Nitrapyrin***–used to inhibit the nitrification process.
- ***Urease inhibitors***–used to slow down the conversion of urea to ammonium and nitrate.
- ***Encapsulation of urea granules***–used to slow the solubility of urea and its release to the soil solution.

Phosphorus Fertilizers

In fertilizer ingredients, phosphorus is commonly represented as an oxide (P_2O_5). Even though this form does not exist in fertilizer materials, it has become the norm for comparing different P fertilizers. Excess P supply in water bodies contribute to excessive growth of plant materials (such as algae blooms), which then die and are degraded by microbes, depleting oxygen in the water and killing fish, shrimp, and other aquatic life. Soil erosion, runoff, and leaching losses of P from agricultural fields are thought to be a primary cause of hypoxia in many parts of the world. Minimizing such agricultural P losses is a part of nutrient management. P that causes hypoxia can also be found in sewage and industrial effluents. P best management methods are intended to reduce P losses to the environment while also increasing the efficiency with which P is used to grow crops. Fertilizers containing phosphorus (P) could be used;

Water-soluble types (quick-acting): single superphosphate (18-20% P2O5); triple superphosphate (45% P2O5).

Partly water-soluble types (quick- and slow-acting): partly acidulated phosphate rock (23-26% P2O5, at least one-third water-soluble).

***Slow-acting types*:** dicalcium phosphate (citrate-soluble); basic slag (citric acid-soluble).

***Very slow-acting types*:** rock phosphate (finely-powdered soft type, e.g., 30% P2O5), with reactivity indicated by formic acid-solubility; permitted minimum is about one-half of total P2O5 content).

***Multi-nutrient fertilizers containing P*:** like N, P: NP; PK (mixtures very commonly used); NPK (may contain about one-third or more water-soluble P for quick supply and two-thirds slow acting P for continuous supply.

Below are the forms of P fertilizers;

Phosphate rock (PR): P reserves are found in historic marine deposits around the world, and PR must be treated to eliminate other minerals. In some cases, unprocessed phosphate rock can be used as a source of P nutrition, but most of it is treated to make other phosphate fertilizers. When PR is given directly to a developing crop, its water solubility may be insufficient to meet the needs of the crop. If utilized on acidic soils, PR can be a good source of P. By reacting PR with acid, over 90% of the PR utilized today is converted into soluble P fertilizers, making it an agronomically and economically viable crop nutrient source.

***Single superphosphate (SSP)*:** The reaction of rock phosphate with sulphuric acid produces SSP. It was the world's first commercial mineral fertilizer, and it paved the way for the contemporary plant nutrient industry to emerge. Because of its low P content, SSP was originally the most regularly used fertilizer, but alternative P fertilizers have mostly replaced it. The SSP fertilizer contains 7 to 9 percent P (16 to 20 percent P_2O_5), 18 to 21 percent Ca, and 11 to 12 percent S in the following proportions: 7 to 9 percent P (16 to 20 percent P_2O_5), 18 to 21 percent Ca, and 11 to 12 percent S.

Triple superphosphate (TSP) [$Ca(H_2PO_4)_2 \cdot H_2O$] or mono-calcium phosphate: TSP was a prominent P fertilizer in the early 1900s, but it has since been phased out in favor of alternative P fertilizers. It has the greatest P content of all the dry fertilizers that don't contain nitrogen, and the P is more than 90% water soluble. It's still popular for legume crops that don't require nitrogen fertilizer.

Monoammonium phosphate ($NH_4H_2PO_4$): Among solid fertilizers, this is the most concentrated source of P. It is most typically made as 11-52-0 and comprises 10 to 12 percent nitrogen and 48 to 61 percent P_2O_5. It's possible to make it with lower-quality phosphoric acid than is used to make other P fertilizers. Monoammonium phosphate is highly soluble and readily accessible to plants in the soil solution as NH_4^+ and $H_2PO_4^-$. MAP can be powdered (typically 61 percent P_2O_5) and used in suspension or clear liquid fertilizers, or sprayed as a foliar spray or added to irrigation water when prepared with purer forms of phosphoric acid.

Diammonium phosphate [$(NH_4)_2HPO_4$] (DAP): The most extensively used P fertilizer in the planet is DAP. It's made by combining ammonia and phosphoric acid. DAP has a standard grade of 18-46-0. It's popular because it contains a high concentration of two regularly used fertilizer components, as well as qualities that make it simple to handle and store. DAP was initially introduced in the 1960s. Because of its high solubility, nutrients are readily available to crops. Because the high ammonium-N concentration of the fertilizer granules can harm seeds and roots near them, it's best to apply it in a band about 10 cm from the seed row, or broadcast and integrate it to prevent concentrating the nutrients too close to the seed or early roots.

***Polyphosphate*:** Polyphosphate is a common liquid phosphate fertilizer made by combining ammonia with phosphoric acid, removing the water, and chaining the phosphate ions together. Single phosphate ions (orthophosphate) can form chains of various lengths, but they are all referred to as "polyphosphate." These fertilizers, which are most typically generated as 10-34-0 or 11-37-0, form transparent liquids that remain stable and crystal-free under a variety of circumstances. As a result, they are a common P source all over the world. Between 25% and 50% of the P in polyphosphate fertilizers is in the orthophosphate (single molecule) form, which is easily absorbed by plants. The remaining 25 to 75 percent of P is in polymers of various lengths that must be broken down in the soil by enzymes or organisms before it can be used by plants. Polyphosphate fertilizers provide a high nutrient content in a clear, crystal-free fluid that is stable over a wide temperature range and has a long storage life. Polyphosphates are a common transporter for micronutrients and other compounds, and they help to ensure consistent dispersion.

Phosphorous Use Efficiency

In soil tests, the solubility of P in specific extractants (water, citric acid, formic acid) is used to determine the rate of transformation under various soil conditions. Water-soluble P (e.g., monocalcium phosphate) is readily available to plants and remains so following immobilization into other forms, though to a lesser level. Granulation and fertilizer placement both slow down this process. Except when fast action is necessary, citrate or citric acid-soluble P is available to plants and can be used for a variety of purposes in a wide range of acidic to neutral soil conditions. Soft powdery rock phosphate contains formic acid-soluble P, which is only available to plants very slowly; its reactivity (release of soluble P) is slightly better where soils are warmer, moister, and more acidic, but still above the acidity damage range. Common P fertilizers produce almost the same yield response per unit of "available" P_2O_5 under intensive farming circumstances on well-fertilized soils. In poor soils, water-soluble P, on the other hand, is better for crops with a short growth season and a limited root system. The acidification from N helps prevent the P from becoming fixed in unavailable forms; gaseous losses of N may occur from surface applied diammonium phosphate (DAP) on neutral soils. Applying N and P fertilizers together in a band or as a combined product sometimes offers advantages for nutrient utilization–the acidification from N helps prevent the P from becoming fixed in unavailable forms; surface applied diammonium phosphate (DAP) on neutral soils may cause gaseous losses of N. Phosphate fixation, or the conversion of soluble fertilizer P into inaccessible forms, is luckily limited to specific soil conditions, such as a high level of active Al and Fe, or a high content of Ca and Mg, as specified by soil pH. P fertilizer consumption is typically around 15% in the first year, but approximately 1-2 percent per year after that, resulting in only about two-thirds being used by the end of thirty years. Weather conditions, soil pH, crop type, and timing and placement of P fertilizer application all affect the effectiveness of P fertilizer use. Fertilizer conducts a series of processes in the soil before being changed to a form that is useful to plants (inorganic phosphate). Because most modern P fertilizers have been treated with sulphuric or phosphoric acid to promote solubility, they are easily soluble. Special treatments can be applied to improve solubility and absorption in some cases, or to minimize fixation into insoluble molecules in others. P can be locked up in insoluble iron or calcium phosphates in very low or very high soil pH circumstances, for example. Organic P sources are insoluble, requiring microbial action to convert the P to an inorganic, usable form. Controlling microbial activity, as with N, can help manage organic P emission.

Potassium Fertilizers

All living cells contain potassium. It is found in relatively tiny levels as the positively charged K+ cation in soil solution and is taken up by plants in that form. The oxide form of potassium fertilizer is the most common (K_2O). This form, like P, is a comparison standard among K fertilizers, but it is not actually found in K fertilizer ingredients. Potash fertilizers are mostly made from saline geological formations. Although low-grade, unprocessed minerals can be employed directly, many fertilizer products now in use are high-concentration, water-soluble, and fast-acting compounds. The following are some examples of K fertilizers:

Potassium chloride (KCl) (0-0-60) or Muriate of Potash (MOP): Most K deposits are found in ancient marine deposits buried deep beneath the Earth's surface as KCl (sylvite) combined with NaCl (halite) in the mineral sylvanite. The ore is crushed, and the KCl and NaCl are separated throughout the processing. The ore is dissolved in hot water and pumped to the surface as soluble sylvanite in a few places, after which the water is evaporated. Solar evaporation is used to recover K salts from saline water in the Dead Sea (Israel/Jordan) and Great Salt Lake (Utah, US). K_2O makes up 60 to 63 percent of KCl (50 to 52 percent K and 45 to 47 percent Cl). It's frequently put on the surface before plowing or banded along the seed row. KCl should not be applied directly to the seed due to its high salt content. It quickly degrades into K+ and Cl- in the soil solution. The potassium binds to cation exchange sites in the clay and organic matter of the soil. The majority of KCl fertilizers are white, however due to the presence of trace amounts of iron oxide, certain K materials are reddish in color; however, both are identical for agronomic application. Pure versions of KCl can be dissolved and used in liquid fertilizers or irrigation water.

Potassium sulphate (K_2SO_4): Potassium sulphate, often known as sulphate of potash (SOP), has a K2O content of 48 to 53 percent and a S content of 17 to 18 percent. Potassium sulphate is found mixed with other minerals in mineral formations. By rinsing the components with water, the components are separated. SOP's K acts similarly to KCl's, but it's also a good supplier of S, in soils that are lacking in it. Because potassium sulphate is less soluble than potassium chloride, it is rarely utilized in irrigation water. If both K and S are required, potassium sulphate is occasionally sprayed as a foliar spray. It's also utilized to provide K to crops that are sensitive to Cl, such as tobacco and potatoes.

Potassium magnesium sulphate (K_2SO_4·2MgSO$_4$): Langbeinite, sulphate of potash magnesia, or commercially Sulpomag are all names for potassium magnesium sulphate. Langbeinite is a rare mineral that can only be found in a few places on the planet. It's sourced commercially from underground mines at Carlsbad, Germany, and Carlsbad, New Mexico (US). Langbeinite has a K_2O content of 21-22 percent, a Mg content of 10-11 percent, and a S content of 21-22 percent. It is a widely used fertilizer in areas where the three key nutrients (K, Mg, and S) are required. It is water soluble but takes a long time to dissolve, and unlike other magnesium and sulfur fertilizers, it has no influence on soil pH.

***Potassium nitrate (KNO_3) or Saltpeter*:** It's a popular fertilizer for high-value crops that require the nitrate form of nitrogen as well as potassium. It's particularly popular as a K source for Cl-sensitive crops. It contains 13 percent nitrogen and 44-46 percent potassium dioxide. It is a common nitrogen source for fertigation and can be applied to the soil or as a foliar treatment to boost fruit development when root activity is diminishing. Several industrial leftovers containing K, such as filter dust, have been created as slower-acting forms, particularly where loss through leaching is required. Fertilizers containing potassium should be used at the time of sowing. Because the K+ ions are adsorbed in the soil, they remain accessible while still being shielded from leaching. Split application, on the other hand, is recommended for crops in soils and climates where larger leaching losses are expected. Although some immobilization into clay lattice layers limits availability, significant fixation into fully inaccessible forms is limited to only a few soil types. During the year of application, the utilization rate of K in fertilizers is around 50-60%.

Sulphur Fertilizers

Inorganic sulphates and organic Sulphur compounds are the most common forms of sulphur in soil. To be taken up by plants, it must be mineralized to the sulphate anion (SO_4^{2-}). In locations where Sulphur-containing fossil fuels are burned, atmospheric Sulphur in the form of Sulphur dioxide

(SO_2) supplies considerable amounts of Sulphur (20 kg/ha or more), but environmental clean-up from such sources has led to a greater requirement for Sulphur fertilization in recent years. Similarly, Sulphur was once delivered to crops through fertilizers such as single super phosphate and various herbicides, but changes in production procedures have diminished those supplies as well. Sulphur in the sulphate form in soils can drain out of the root zone, particularly in coarse-textured soils. Irrigation water may be a key source of S for crops in irrigated agricultural systems. For formulating recommendations, the soil tests available for S are not very dependable. A variety of fertilizer products can be used to supply sulphur to field crops.

Elemental sulphur (0-0-0-90 S): Elemental sulphur is a practical form of sulphur that may be given to a variety of crops in a broadcast or banded application. Physical techniques must be used to break down elemental sulphur into small particles before it can be applied to the soil. Thiobacillus and other soil bacteria convert it to sulphate ions once it reaches the soil. Sulphur can be broken down into small particles more quickly by combining it with bentonite clay during the formulation process. A bentonite-sulphur particle swells in water, breaking up into extremely small particles. The increased surface area allows soil bacteria to convert sulphur to sulphate more quickly after it is broken down into minute particles. Even in the presence of tiny particles, however, the change of sulphur to sulphate is a laborious and often months-long process. As a result, rather than using elemental sulphur, a sulphate fertilizer like ammonium sulphate is recommended for most crops in the first sulphur fertilization year.

Ammonium sulphate (21-0-0-24 S): Ammonium sulphate is a byproduct of the chemical industry that can be used as a N source when S is also required. Because it is more accessible to plants and less subject to leaching losses than many other sources, ammonium sulphate is a preferred source of S. It also offers the advantage of meeting a portion of the crop's nitrogen requirements.

Calcium sulphate (Gypsum) (24% S): For many years, gypsum has been widely utilized for fertilizing and soil restoration as a sulphur and calcium-bearing substance. It's a neutral salt that has no influence on the acidity of the soil. Gypsum is less soluble than ammonium sulphate and more difficult to work with.

Single superphosphate (SSP) (0-20-0); (8 to 10% S): Sulphur deficiency is uncommon on ground that has been appropriately fertilized to feed P to crops via SSP. However, in the production of concentrated superphosphates like triple superphosphate, the gypsum is substantially eliminated, leaving these compounds with little or no sulphur.

Ammonium polysulfide (45% S): Ammonium polysulfide can be sprayed directly to the soil, combined with anhydrous ammonia or ammonia solutions, or added to irrigation water. Ammonia polysulfide, on the other hand, is not totally compatible with very acidic or high-salt liquid fertilizers. In some cases, it's used to lower pH and improve water infiltration into the soil.

Ammonium thiosulphate (26% S): Another sulphur-containing substance is ammonium thiosulphate. It's safe to use in irrigation water. Many fertilizer solutions, such as aqua ammonia, N solutions containing ammonium nitrate, urea solutions, and most N, NP, or complete fertilizer solutions, are also compatible with it. It cannot, however, be mixed with anhydrous ammonia or acid solutions like phosphoric acid, since the thiosulphates will be decomposed.

Calcium

Depending on the mineralogy, calcium makes up anywhere from 0.1 to 25% of soil. Liming to keep the pH in the right range for plant growth usually provides enough calcium to suit agricultural needs. Ca is an important component of plant cell walls. Under low pH, the toxicity of other nutrients (such as aluminum, manganese, and copper) might mask Ca deficiency in plants. When there is a requirement to remedy a Ca deficiency yet the pH is high, gypsum might be utilized to deliver Ca.

Magnesium

Magnesium is a constituent of chlorophyll. As a result, Mg shortage frequently manifests as yellowing between the veins of leaves. The most common method of addressing Mg deficiency is to use dolomitic lime (11 percent Mg). In acid soils or areas where high rainfall causes Mg leaching from the root zone, more soluble forms of Mg (such as $MgSO_4$) may be required to remedy the deficit quickly. Magnesium fertilizers are either fast-acting soluble salts (such as sulphates) or slow-acting soluble salts (such as nitrates) (such as dolomitic lime). Magnesium sulphate, which comes in the form of Epsom salts (10 percent Mg) or keiserite (16 percent Mg), is the most common Mg fertilizer for crops. Keiserite, also known as magnesium sulphate monohydrate ($MgSO_4H_2O$), is a natural mineral found in Germany's deep subsurface geologic marine deposits. For plant nutrition, it supplies a soluble source of both Mg and S. The fine crystalline kieserite can be applied directly to the soil, or it can be granulated into a larger particle size for mechanical fertilizer distribution or bulk combining with other fertilizers. In instances where K is also insufficient, potassium magnesium sulphate (see potash fertilizers) is an essential source of Mg. Magnesium carbonate is a mineral that contains magnesium (dolomitic lime). Dolomitic lime or magnesium carbonate are the cheapest sources of Mg for liming soils. A high rate of dolomitic lime application can assure a steady supply of Mg for several years with no negative consequences.

Enhancing Fertilizer use Efficiency

Fertilizers can be made more efficient by slowing nutrient release, preventing conversion to less stable forms in the soil, or increasing nutrient availability to plants. The decision to use these changes to improve fertilizer efficiency is based on prospective agronomic, economic, and environmental benefits. Fertilizer formulations and additions that serve to boost the efficiency of the applied fertilizer have sparked increased interest in recent years. To meet this need, a variety of procedures and products are available. To keep fertilizer granules from dissolving, one option is to establish a physical barrier or coating. The protection can last anywhere from a few days to a few months, depending on the thickness of the coating and its components. Another option is to create a chemical mixture that boosts nutrient absorption. There are several products on the market that disrupt the biological processes of nutrients in the soil. These include nitrification inhibitors, urease inhibitors, and other compounds that can be employed to halt or slow down soil bacteria-controlled conversion pathways.

Coatings and controlled-release fertilizers: Coatings on soluble fertilizers have been made from a variety of materials. Coatings are most typically used on granular or prilled N fertilizers, although they are also used on multi-nutrient fertilizers. Urea is the most often coated fertilizer since it has the largest nitrogen content of all conventional soluble fertilizers. The first extensively used fertilizer coating was Elemental S. It entailed spraying molten S over urea granules, then applying sealant wax to fill in any gaps or defects in the covering. Later, when the S layer was covered with a thin layer of organic polymer, the procedure was improved. To limit nutrient release and improve nutrient use efficiency, polymer coatings can be added to fertilizer particles. Depending on the coating and the conditions in the soil, these coatings can protect fertilizer particles for a few weeks to several months. For a limited time, certain coatings prevent fertilizer particles from decomposing in the soil. The ecologically smart N coating is an example of a coating that slows the release of nitrogen in urea (ESN). It consists of granular urea encased in a flexible polymer covering that regulates water absorption and dissolved nitrogen release. The polymer coating can be applied at various thicknesses to control the rate of water absorption and urea dissolution. Coated products work by storing nutrients in a form that is less likely to be lost from the root zone and releasing them at a time that is as close to the plant's needed absorption time as possible. Even a few days of delayed release can result in large nutrient losses and, as a result, a significant increase in nutrient utilization efficiency. Losses to the environment are also avoided by keeping more of the nutrient in the crop.

***Inhibitors*:** There have been a variety of chemical and biological inhibitors discovered that can be employed to control nutrient release or activity in the soil. These chemicals are designed to prevent a certain reaction in the soil from occurring for a set length of time. Nitrogen is the most targeted nutrient once again. Nitrification inhibitors, which limit the bacterial conversion of ammonium to nitrate, and urease inhibitors, which impede the enzymatic breakdown of urea to ammonium, are two examples of goods. Additions of urease and nitrification inhibitors to urea allow applied N fertilizer to stay in the soil closer to the root zone for longer periods of time, allowing for improved plant uptake and reducing possible N losses from the soil-plant system to the environment. Several soil and environmental conditions, however, influence the efficiency of these inhibitors in raising yield and improving grain protein content. Thus, for various crops, soils, and climates, the best application technique and timing of N fertilizer when combined with inhibitors must be determined.

***Others*:** Some micronutrients can be integrated into glass beads, making it easier to distribute the nutrients uniformly and controlling the rate of release. In some cases, liquid polymers are employed to bind with soil cations and aid maintain P solubility. To address specific nutrient management requirements, a variety of alternatives are available. Adding these control systems typically raises expenses, but they may be justified if nutrient efficiency and yield are improved sufficiently.

GAP for Fertilizers

Though fertilizers are the most common source of plant nutrients, they also have the potential to harm soil characteristics, pollute soil and water, and reduce fertilizer efficiency if they are not carefully chosen based on soil conditions and features. The cost of fertilizers is determined by the method and timing of application.

- Nitrate nitrogen fertilizers outperform other types of nitrogen fertilizers under dry soil conditions.
- Continued use of ammonium sulphate, urea, ammonium chloride, and ammonium sulphate on acidic or low-lime soils
- Nitrate should be avoided.
- Because all nitrogenous fertilizers are equally effective and nitrogen loss due to leaching is uncommon during the Rabi season, nitrogenous fertilizers should be chosen based on cost per unit weight of nitrogen rather than other criteria.
- Nitrogenous fertilizers are suitable for broadcasting and top-dressing due to their solubility and mobility, whereas phosphatic and potassic fertilizers should be put near the seed or root zone.
- Because plants require nitrogen throughout their growth phase and nitrogen fertilizer is lost through leaching and other processes, a single nitrogen application should be avoided and nitrogen should be provided in divided doses during the growth period.
- Because plants require a large amount of phosphorus during the early growth stage, and because all phosphatic fertilizer becomes available to the plant slowly, the complete dose of required phosphatic fertilizer should be applied on or before sowing.
- To ensure consistent availability of potash, the entire amount of potassic fertilizers should be applied at the time of planting. On sandy soils, splitting potassium is desired to improve use efficiency.
- Fertilizer should not be used during periods of heavy rain.

Organic Fertilizers

A variety of organic materials can serve as soil amendments and nutrition sources. Because many are waste products, they can be inexpensive, especially if used close to where they are manufactured. Some farm wastes are recycled because it is the only and most cost-effective way to dispose of them. If a farmer needs purchase waste material of any kind, it must be relatively inexpensive, have no harmful or hazardous effects, and be profitable. Organic fertilizers normally necessitate handling and transporting significant volumes of material in order to acquire relatively low levels of nutrients; hence they should be used on areas close to the source. Processed organic wastes, especially if sold, require mechanical and chemical preparation, such as drying, grinding, mixing, granulating, neutralizing, supplementing with specific nutrients, and being free of pathogenic microorganisms. The following are important characteristics for organic fertilizers:

- dry matter content,
- total and easily mineralizable organic matter,
- total and quick-acting N,
- C/N ratio,
- total P and K contents,
- content of substances detrimental to plant growth or product quality (heavy metals should be below established critical limits).

The rates at which organic materials are applied should consider both the expected nutrient supply available to the crop (for N in farmyard manure, about 20-30% of the amount applied is available to the crop during the first year; this can be relatively high in tropical climates) and the need to minimize nutrient losses or other negative effects. Nutrients from organic sources have the same nutritional value as those from mineral fertilizers. There is no scientific evidence that organic nutrition sources are healthier for plants. Organic materials, in addition to providing nutrients, may have other advantages for plant growth, such as increasing the physical and chemical qualities of the soil. Organic fertilizers come in a variety of forms.

- ***Naturally occurring material:*** *(e.g., peat)*
- ***Farm wastes*****:** crop leftovers (straw, leaves, etc.); animal manures (farmyard manure, liquid manure, slurry); compost (compost mixture of decomposed plant residues); green manures (compost combination of decomposed plant residues); (leguminous or other crops incorporated into the soil).
- ***Residues from processing of plant products, e.g.*****:** pressed cakes (from oilseeds) and fibers (from the paper business); wood materials (bark, sawdust, lignin from the paper industry); molasses (from the sugar industry); seaweed extracts
- ***Residues from processing of animal products, e.g.*****:** blood-, horn- and bone-meal; leather dust, etc.
- ***Urban wastes:*** composted household refuse; sewage sludge.

- ***Soil inoculants***: (e.g., Living micro-organisms).

Efficiency-enhancing additives can be added to manures and other organic nutrient sources, just like mineral fertilizer components, to reduce the risk of nutrient losses during storage or after application to the field. Options for placement and timing may also help with organic nutrient control. Most organic matter accumulation is reliant on plant development, which is reliant on nutrient balance in the soil. Organic matter derived from fertile soil can store nutrients in the soil until they are transformed to plant-available forms, providing a slow-release mechanism for feeding nutrients to the following crop. As a result, fixing nutrient shortages is far more significant than increasing soil organic matter in terms of enhancing soil productivity.

Microbial Inoculant

Microbial inoculants, also known as soil inoculants or bioinoculants, are agricultural amendments that boost plant health by utilizing beneficial rhizospheres or endophytic bacteria. Many of the microorganisms involved create beneficial symbiotic interactions with the target crops (mutualism). Microbial inoculants can be used to increase plant nutrition as well as promote plant growth by boosting the generation of plant hormones. The benefits of inoculants in agriculture go beyond their use as biofertilizers, according to research. Microbial inoculants can cause crop species to develop systemic acquired resistance (SAR) to a variety of common crop diseases (provides resistance against pathogens). Furthermore, it is becoming more widely understood that microbial inoculants frequently alter the soil microbial community, which should be considered. Microbial inoculants come in a variety of forms.

Rhizobacterial inoculants: Nitrogen-fixers, phosphate-solubilizers, and other root-associated beneficial bacteria are among the rhizobacteria widely used as inoculants. These bacteria help the host plant get more nitrogen and phosphorous from the macronutrients. Plant growth-promoting rhizobacteria are the common name for these bacteria (PGPR). Among them are:

- ***Nitrogen-fixing bacteria***: Rhizobium and closely related genera are the most widely used rhizobacteria. Rhizobium are nitrogen-fixing bacteria that live in nodules on legume roots and create symbiotic relationships with them. This improves the nitrogen nutrition of the host and is critical to produce soybeans, chickpeas, and other leguminous crops. Azospirillum has been shown to be useful for nitrogen fixation and plant nutrition in non-leguminous crops in some circumstances. Diazotrophic rhizobacteria have improved plant growth, grain yield, nitrogen, and phosphorus uptake in cereal crops. Rhizobacteria are bacteria that dwell in root nodes and are linked to legumes.

- ***Phosphate-solubilizing bacteria*:** The use of phosphate-solubilizing bacteria (PSB) such as Agrobacterium radio-bacter to increase phosphorus feeding has also gained interest. PSB bacteria are free-living bacteria that break down inorganic soil phosphates into simpler forms that plants can absorb.

***Fungal inoculants*:** A Mycorrhiza connection is a symbiotic link between fungi and plant roots. This symbiotic relationship can be found in nearly all land plants, and it benefits both the plant and the fungus. In exchange for enlarging the root absorptive area with hyphae, the plant can provide up to 30% of its energy output to the fungi, giving the plant access to nutrients it would otherwise be unable to obtain. Arbuscular mycorrhizae and *ectomycorrhizae* are the two most frequent mycorrhizae. *Ectomycorrhizae* connections are most abundant in woody species and have fewer consequences for agricultural systems. These are some of them:

***Arbuscular mycorrhiza*:** Because of its ability to access and give phosphorus to the host plant, arbuscular mycorrhiza (AM) has been considered as a viable agriculture modification. Tomato yields of 100 percent fertility were achieved at 70 percent fertility using a reduced fertilization greenhouse system infected with a mixture of AM fungus and rhizobacteria. This 30% reduction in fertilizer application can help reduce nutrient contamination and extend the life of finite mineral resources like phosphorus (Peak phosphorus). Increases in salinity tolerance, drought tolerance, and resistance to trace metal toxicity are among the other impacts.

Fungal partners: Host plants can benefit from fungus inoculation alone. Inoculation along with additional changes can help to improve the situation even further. A frequent household additive for personal gardens, agriculture, and nurseries is arbuscular mycorrhizal inoculation mixed with compost. This coupling has also been shown to promote microbial functions in soils that have been impacted by mining. In various ecotones or with certain crops, certain fungal partners work well. In highland rice paddies, arbuscular mycorrhizal inoculation combined with plant growth boosting bacteria resulted in a higher yield and faster maturation. The addition of arbuscular mycorrhizae and biochar to maize increased its growth. This modification can also help crops absorb less cadmium.

Approaches to Fertilizer Recommendations

Fertilizer recommendations for crops are divided into three categories. The sufficiency approach is concerned with the demands of the plant, whereas the build-up and maintenance approaches are concerned with the amount of nutrient available in the soil. Because the amount extracted by the plant is restricted by the concentration at the root-soil interface, the uptake of immobile nutrients (such as P and K) by crop plants is a function of the concentration of plant accessible nutrients in the soil. When a nutrient is limited to that level, percent sufficiency is calculated using the soil test value expressed as a percentage of the potential yield. Each nutrient has a suggested application rate and a sufficiency index for each crop. The sufficiency strategy aims to apply just enough fertilizer to maximize profits in the current year of application while reducing nutrient applications and fertilizer expenditures. For efficient fertilizer management, the sufficiency technique is usually referred to as the "feed-the-crop" strategy. To determine sufficiency levels, the soil test levels must be calibrated using yield response trials. The critical level is determined in these trials as the threshold at which yield does not rise. The sufficiency approach has the advantage of maximizing output while reducing annual inputs. To sustain those yield levels, though, applications will need to be submitted every year. Short-term leases are well-suited to fertilization based on sufficiency levels.

***Maintenance approach*:** Nutrients that were withdrawn with the crop at harvest should be replenished using the maintenance method to nutrient management. To maintain the soil nutrient level, fertilizer is applied based on the amount of nutrients withdrawn from the field. When soil nutrient levels are over the critical threshold, the maintenance concept does not propose application. The soil will be able to supply the nutrients required by the crop above the critical soil test level, and no fertilizer reaction is expected. The maintenance concept assumes that the crop nutrient removal rate is accurate, allowing soil testing to remain at crucial levels. The actual amount needed for maintenance may be slightly higher than the amount removed by the harvested crop, to account for part of the applied nutrients being held up in chemical interactions in the soil.

***Build-up approach*:** The "feed-the-soil" rather than "feed-the-plant" approach to nutrient management is used in the build-up technique. To get the concentrations to the point where they won't be limiting, more nutrients are applied than are withdrawn by the crop. This strategy allows you to take advantage of the exceptional 'good years,' when weather and other factors contribute to higher-than-average yields. Depending on the farmer's financial status, the process of accumulating soil test values normally takes four to eight years. Longer-term build-up schemes aid farmers in budgeting by spreading the cost of fertilizer across several years. Shorter build-up programs, on the other hand, can deliver benefits from higher soil testing sooner. Due to severe leaching of the nutrients given or the nutrients being 'tied-up' in inaccessible forms, building soil test levels is not possible in some soils. Such restrictions are common in soils with high sand concentration or high organic matter levels. Fertilizer should be applied as soon to the period of crop usage as possible in these circumstances to get the most efficient use of applied nutrients.

***Build-up and maintenance approach*:** The terms "build-up" and "maintenance" are frequently used interchangeably. The build-up and maintenance philosophy states that fertilizer is administered over a specified time period until nutrient levels (such as P and K) are raised to critical soil test levels, and then applications are sustained at a rate to keep the nutrient levels at the critical soil test levels. This method necessitates the use of a calibration data set that depicts the link between the amount of fertilizer added and the change in soil test level. For a given soil test process, the soil test yields an indication of the soil's productivity at various soil test levels. The build-up and maintenance method leads to soil test levels in the range where a yield response to added fertilizer is unlikely. To sustain optimum yields and ensure that nutrients are not restricting, soil test levels are maintained. This method is usually effective for P and K, but it is ineffective for N since N soil testing cannot be built up or sustained. The construction of a critical level is the most significant aspect of this paradigm. This crucial threshold is the same as the 100 percent level set for the sufficiency concept. The fundamental distinction is that the sufficiency strategy aims to apply fertilizer only up to the critical value. The purpose of the build-up and maintenance technique is to apply fertilizer until no more crop response is predicted from more fertilizer application, then try to keep the soil test at that level. With the sufficiency strategy, fertilizer must be applied on an annual basis to attain the critical level required to maintain output. Because the crop is less reliant on the annual spray, the maintenance technique provides greater flexibility. So, if weather or economics force a reduction in fertilizer use for one crop season, yield is unaffected. When the soil test exceeds the sufficiency approach's "critical level" or the build and maintenance approach's "maintenance limit," the recommendation is to apply only starter fertilizer or none until the next testing cycle. The calibration data used to relate the laboratory data to crop yield is based on extensive field studies for each crop and different soils, and the interpretation of soil tests is dependent on the laboratory procedures used and the calibration data used to relate the laboratory data to crop yield is based on extensive field studies for each crop and different soils. On neutral-to-acid soils, the Bray P1 or Mehlich-3 colorimetric techniques, or the Mehlich-3 procedure on soils with pH > 7.3, are used to determine the relationship between predicted yield and soil P. These values should not be utilized for the Olsen (soil bicarbonate) test or inductively coupled plasma emission spectroscopy analysis of Mehlich-3 extractions (ICP). Soil test findings are an index value that must be matched to the necessary calibration data tying the lab measurement to relative yield values, rather than an actual assessment of nutrient levels in the soil.

Fertilizer Application Methods

When it comes to fertilizer application methods, there are various aspects to consider. The specific combination of nutrients required, the crop, treatment time, available equipment, and planting/tillage procedures are all factors to consider. In the table below, each application method is detailed along with its benefits and drawbacks.

Table 21: Fertilizer Application Method and their Characters

Application method	**Characteristics**
Banding	
Small amounts of liquid or granular fertilizer are applied in furrows 1-2 inches below and 2-3 inches to the side of seed placement	• Liquid or dry fertilizer • Nutrient uptake efficiency is higher than for broadcast application, allowing for lower rates • Runoff risk is low • Equipment is standard on many planters • Fertilizer placement in vicinity of roots may aid root development • Requires precision placement to avoid placement too close to the seed
Furrow/pop-up/starter application	

Application method used alone or combined with banded application. Usually applied in low rates placed directly on or around the seed during planting.	• Liquid fertilizers are best suited • Can be applied in combination with banded fertilizer during planting • Only low rates needed • Can only be used in low rates to avoid phytotoxicity
Broadcast	
Spinning of dry granules or spraying of liquid fertilizer onto the soil surface, with or without incorporation of the material	• Can be done quickly • Standard equipment used • Depending on the source, it requires rainfall or tillage to move fertilizers to roots • Could result in poor application uniformity
Topdressing	
Spreading or broadcasting of fertilizers in the standing crop, after emergence.	• Fast and easy application of fertilizer • Can result in volatilization losses for urea containing sources • Could cause leaf burn • Could result in poor application uniformity
Side-dressing	
Broadcast or banded between or alongside growing plants; this method normally refers to N applied when corn is 6-12 inches tall	• N can be applied with a high nutrient use efficiency • Application with drop nozzles can be done faster than injection or knifing • Weather and field conditions can delay or prevent application
Foliar feeding	
Application where liquid fertilizer is applied directly to the leaves, nutrients are in highly available form and can be taken up by the plant rapidly	• Easy application • Standard sprayer equipment can be used • Not always practical due to row spacing • Material must be absorbed by leaves before it dries

Source: http://nmsp.cals.cornell.edu/guidelines/factsheets75.html

Chapter 5

Soil Amendments

CONTENTS

INTRODUCTION

Any material added to a soil to improve physical attributes such as water retention, permeability, water infiltration, drainage, aeration, and structure is referred to as a soil amendment. The purpose is to improve the root system's environment. An amendment must be completely incorporated into the soil before it can act. Its efficiency is lessened if it is simply buried, and it will obstruct water and air flow, as well as root growth. Although many mulches are also used as amendments, amending a soil is not the same as mulching. On top of the soil, a layer of mulch is applied. Its goal is to reduce evaporation and runoff, as well as to prevent weed development and provide a pleasing appearance. Mulches also help to keep the soil temperature down. After organic mulches have degraded to the point where they no longer serve their purpose, they can be integrated into the soil as amendments. The purpose of soil amendment is to improve the root and plant growth environment, which includes improving the soil structure and water holding capacity, nutrient availability, and the living conditions for soil organisms, which are necessary for plant growth. Additionally, improved soil texture and root growth prevent soil degradation during heavy rains or in windy areas. When organic amendments are utilized, it also helps to support the nutrition cycle (e.g., manure). Aside from soil amendment, there are a variety of ways to conserve soil moisture, including soil cover and reforestation (living plants), mulching, and various tillage techniques. Of course, it is also critical to plant a crop that is suited to the local climate. Soil amendments are basically any organic or inorganic material that is put to the soil to improve its condition. The type of amendment to use is entirely determined by how the soil needs to be altered. Almost every type of soil can be made fruitful by adding soil amendments. The soil can be amended with both inorganic and organic materials. Inorganic substances are mined or man-made, whereas organic substances are derived from living creatures (e.g., plants).

***Inorganic Amendments*:** Vermiculite, perlite, tire pieces, pea gravel, and sand are examples of inorganic materials. They must be purchased in most cases, making them more expensive than organic amendments. Furthermore, the industrial manufacturing process necessitates a significant amount of energy. As a result, they do not have the same level of long-term viability as organic amendments. Most are sterile (as far as plant pathogens are concerned) and many are inactive.

***Organic Amendments*:** Organic growing medium amendments are typically derived from naturally occurring plants or plant products (peat moss from peat bogs) or by-products of processing plants or mills (sawdust, cedar chips, bark, bagasse, rice hulls) or waste disposal plants (sawdust, cedar chips, bark, bagasse, rice hulls) (compost, processed sewage sludge, biosolids). Compost and animal dung are by far the most important organic soil additives on the planet (e.g., chicken, cow). Compost and animal manure are frequently readily available and inexpensive. Peat moss is the most widely utilized organic amendment in professional agriculture (e.g., sphagnum peat moss which is the highest quality, but also hypanaceous or domestic peat). Wood chips, grass clippings, straw, compost in general (made in large- or small-scale facilities), manure (see also usage of compost, use of dehydrated feces or terra preta), processed sewage sludge (biosolids), sawdust, and other organic supplements are examples. Organic substances are usually more readily available than inorganic substances, and they are generally less expensive. Soil amendment is improved by using excellent crop rotations and selecting the right crop. Most organic amendments, especially sawdust, cedar chips, bark, and bagasse, should be composted or aged before being used in ground beds, and sterilized if possible. The C:N ratio of all these changes is high. Use of non-compostable amendments with a high C:N ratio will deplete N in the soil, may result in a salt or ammonia/ammonium burn, or may result in heat damage.

ADVANTAGES OF ORGANIC AMENDMENTS

Humic acid is a key component of humic substances, which are the most abundant organic component in organic soil amendments. Farmers have known for longer than recorded history that organic soil additives (humus) are advantageous to plant growth. Plants were thought to utilise humus directly, but it was later discovered that plant growth is dependent on inorganic chemicals. As a result, many soil scientists believed that organic matter was only beneficial for fertility when it was broken down and its constituent nutrient elements were released into inorganic forms. Soil scientists now use a more holistic approach, recognizing that humus regulates soil fertility by affecting the soil's water-holding capacity: the spongy structure of organic matter can bind water and some inorganic compounds that act as micro- or macronutrients. As a result, humic acids decrease the evaporation of water from soils. This is especially significant in soils that lack or have a low percentage of clay, in arid locations, and in sandy soils that can't store water. Another water molecule's oxygen end interacts with another's hydrogen end until the evaporation rate is reduced by 30%. Another advantage of organic amendments is that organic matter nourishes soil bacteria, which in turn release nutrients into the soil, boosting soil fertility.

SOIL AMENDMENT PRODUCTS

Wood Products: Wood products can bind nitrogen in the soil, causing plant nitrogen shortage. Nitrogen is used by soil microorganisms to break down the wood. The nitrogen is released after a few months and is once again available to plants. Because sawdust has a larger surface area than wood chips, it poses the greatest risk. To avoid nitrogen deficit, you may need to apply nitrogen fertilizer at the same time as wood chips or sawdust.

Sphagnum Peat vs. Mountain Peat: Sphagnum peat is a good soil amendment, particularly for sandy soils, which retain more water after being amended with sphagnum peat. Sphagnum peat is acidic (low pH) and can assist gardeners in growing plants that demand acidic soil. Mountain peat from Colorado isn't as effective as other soil amendments. It has a delicate texture and a higher pH than most other foods. Mountain peat is extracted from high-altitude wetlands that will take hundreds, if not thousands, of years to regenerate. Mining has a significant impact on hydrologic cycles and mountain ecosystems. Bogs in Canada and the northern United States are used to gather sphagnum peat. After harvest, the bogs can be re-vegetated and grow back swiftly in this damp environment.

Biosolids: Biosolids are sewage treatment byproducts. They can be found on their own or mixed up with leaves or other organic elements in compost. Heavy metal content, pathogen levels, and salts are the main problems with biosolids. Always use a Grade 1 biosolid to avoid excessive amounts of heavy metals and to ensure that microorganisms have been destroyed. While Grade 1 biosolids are suitable for food gardens, they should not be used on root crops because they will come into direct touch with the plant's edible section. Grade 1 biosolids should not be used.

Manure vs. Compost: Because of the high quantities of ammonia in fresh manure, it can injure plants. To avoid this issue, exclusively use older manure (at least six months old). Pathogens are also a concern with fresh manure, particularly in vegetable gardens. Before spreading manure to vegetable plants, compost it for at least two heating cycles at 130 to 140 degrees F to kill any germs. Most home composting systems are unable to maintain temperatures of this magnitude. Manure-based home composted products work best in flower gardens, shrub borders, and other non-food gardens. Ammonia gas is released from the manure during composting. As a result, nitrogen levels in composted manure may be lower than in raw manure. Composted manure, on the other hand, will have higher phosphate and potassium contents. Adjust your fertilizer usage accordingly. Composted manure will also have higher salt levels than raw manure. Do not apply manures to your garden soil if the salt levels are already high.

Additional amendments include:

- Lime - raises soil pH, reducing acidity
- Sulfur - lowers soil pH, increasing acidity
- Gypsum - improves aeration of compacted soil, helping it drain more efficiently
- Perlite - improves aeration and drainage
- Vermiculite - improves moisture retention and aeration

Agricultural lime: A soil addition prepared from pulverized limestone or chalk, agricultural lime is also known as ag-lime, agricultural limestone, garden lime, or liming. Calcium carbonate is the main active ingredient. Calcium oxide is one of the additional compounds that may be present depending on the mineral source. Powdered limestone, unlike quicklime (calcium oxide) and slaked lime (calcium hydroxide), does not require lime to be burned in a kiln; it simply requires milling. All these kinds of lime are occasionally employed as soil conditioners, with the general theme of supplying a base to rectify acidity, but crushed limestone is most commonly utilized in farm fields today. Historically, burnt lime was commonly used to lime agricultural fields; the discrepancy can be explained in part by the fact that affordable mass-production-scale fine milling of stone and ore relies on methods established since the mid-nineteenth century. The following are some of the effects of agricultural lime on soil:

- it increases the pH of acidic soil, reducing soil acidity and increasing alkalinity
- it provides a source of calcium for plants
- it improves water penetration for acidic soils
- it improves the uptake of major plant nutrients (nitrogen, phosphorus, and potassium) of plants growing on acid soils.

Dolomitic lime and hydrated lime are two more types of lime that are commonly used in agriculture and gardening. Dolomitic lime can be used as a soil amendment to have similar effects as agricultural lime while also providing magnesium. Hydrated lime can be used as a disinfectant in animal production, creating a dry, alkaline environment in which bacteria cannot easily reproduce. It can be used as an insect repellent in horticulture farming without harming the pest or the plant. Agricultural lime is typically spread on fields with spinner-style lime spreaders. Agricultural lime is injected into coal-burning power stations to minimize emissions of pollutants including NO2 and SO2. Lime can boost crop productivity and the root system of plants and grass in acidic soils. It accomplishes this by making the soil more basic, allowing the plants to absorb more nutrients. Lime is not a fertilizer in and of itself, although it can be used in conjunction with one. A soil test, which can be provided by a university with an agricultural education department, is the best technique to identify if a soil is acidic or lacking in calcium or magnesium.

There are at least four factors to consider in selecting a soil amendment:

- how long the amendment will last in the soil,
- soil texture,
- soil salinity and plant sensitivities to salts, and
- salt content and pH of the amendment.

Organic additions' salt content, pH, and organic matter can all be determined using laboratory testing. After that, the quality of bulk organic amendments for application in large-scale landscapes can be determined.

***Longevity of the Amendment*:** The amendment you choose depends on your goals.

- Are you looking for a quick way to improve the physical qualities of your soil? Select a decomposer that is quick to decompose.
- Do you want to make a long-term difference in your soil? Select a decomposer that takes a long time to decompose.
- Do you want a quick fix that will last for a long time? Choose from a variety of amendments.
- Soil Texture

***Soil texture*:** The size of the soil particles is reflected in the texture of the soil. Sandy soils are characterized by big soil particles and a gritty texture. Clay soils are sticky and have tiny soil particles. Gardeners have challenges in both sandy and clay soils. The optimal blend of varied size soil particles can be found in loam soils. The purpose of amending sandy soils is to improve the soil's ability to retain moisture and nutrients. Use well-decomposed organic amendments, such as composts or aged manures, to achieve this. The goal with clay soils is to improve aggregation, porosity, and permeability, as well as aeration and drainage. In this case, fibrous amendments like as peat, wood chips, tree bark, or straw are the most beneficial. Because sandy soils retain little water, apply a high-water-retention supplement like peat, compost, or vermiculite. Because clay soils have low permeability, use a permeability-boosting amendment like wood chips, hardwood bark, or perlite. Because of its high-water retention, vermiculite is not a good choice for clay soils.

***Soil Salinity and Plan Sensitivity to Salts*:** Salt levels in some composts and manures can be considerable. Avoid using these amendments in soils with excessive salt levels (more than 3 mmhos/cm) or when growing salt-sensitive plants. Many landscape plants, including raspberry, strawberry, bean, carrot, onion, Kentucky bluegrass, maple, pine, viburnum, and others, are salt sensitive. Instead of compost or manures, use sphagnum peat or ground leaves in certain situations.

***Salt Content and pH*:** The amendment's salt content and pH Salts in soil amendments should always be avoided. In Colorado soils, excessive salinity and pH are common issues. As a result, stay away from amendments that are heavy in salts or have a high pH. Wood ash, Colorado mountain peat, and composted manures are examples of amendments that are rich in salts and/or pH. If thoroughly blended with low-salt soils (less than 1 mmhos/cm), an amendment with up to 10 mmhos/cm total salts is acceptable. Amendments with a salt content of more over 10 mmhos/ cm should be avoided. If your soil has a high salt content, use a low-salt amendment. Sphagnum peat and compost formed entirely of plant matter are low in salts and are excellent amendments for Colorado soils. Request a study of the organic modifications you're considering, and make informed decisions. Test a little amount of the modification before purchasing a large quantity if no analysis is provided.

Other factors may include:

- ***Cost Considerations*:** Most inorganic amendments are mined from natural deposits and processed further to produce the finished product. To make such adjustments, you'll need a lot of heavy machinery and a lot of technical know-how. As a result of these realities, production costs have increased. Organic amendments, on the other hand, have a high self-help compatibility since farmers may make them on-site. There are a variety of strategies for producing organic amendments and ensuring a long-term management of the local nutrient and water cycle. Composting toilets, also known as terra preta toilets, fossa alternas, or arborloos, and UDDT, are toilets that produce organic amendments. Organic compounds are more plentiful than inorganic substances, and they are often less expensive.

- ***Environmental Effects*:** Soil amendments must be treated with caution. The potential of soil to bind chemicals indefinitely without releasing them is limited. When too much nutrients or organic matter are added to the soil, they can be discharged into the groundwater and adjacent rivers and lakes, resulting in water pollution.
- ***Applicability*:** Soil amendments can be used anywhere if they are properly handled. Depending on the initial soil conditions, care must be made while selecting the proper soil amendment. Organic additions can be made locally. It can take a long time to mix the mixture into the soil.

Chapter 6

Soil Testing

CONTENTS

INTRODUCTION

The most prevalent nutrient management diagnostic method is soil testing. A competent soil testing program can assist in determining the nutrients available in the soil, the requirement for supplemental build-up treatments, and trends in nutrient levels over time as a result of crop removal, fertilizer addition, and other factors. It's crucial to remember that most soil tests are an index of the soil supply rather than a measurement of the nutrients in the soil. As a result, soil test numbers are only meaningful if they are linked to a set of calibration data that shows the relationship between the soil test index number and the likelihood of a response to additional nutrients. Scientists have devised several soil testing methodologies to determine the amount of nutrients in the soil as well as the availability of key plant nutrients to developing crops. The soil test must be calibrated through rate studies to establish an indicator of the plant accessible nutrient supply because the actual amounts of each nutrient in the soil are far more than what is available to the growing crop. To account for responses to fertilizer additions in diverse soil types and climate locations, calibration data is required. The calibration data used to interpret soil testing should come from the same soil type and environment as the field for which the recommendations are being made. For general recommendations, official university and government calibrations are available, but as we try to fine-tune nutrient management for specific field characteristics, it's critical to recalibrate soil test recommendations to the specific soils, climate, and management practices used on the farm. As a result, having local calibration data to utilize in interpreting soil test results for a given field becomes important. More precision in soil testing and advice is being requested by farmers and their advisors. The generic advice of the past is no longer sufficient. When it comes to soil nutrients, a "checkbook balancing" approach isn't applicable. The availability of nutrients applied and the impact of nutrients removed are affected by a variety of efficiency parameters. The soil has a significant buffering capacity that has an impact on nutrient balance. One of the most important aspects of soil testing is to remember that the sample taken is only as good as the test. To best represent the area being sampled, samples must be collected carefully and systematically. Sample collecting procedures should be consistent with those used to collect samples for calibrating the soil test. One method is to use a random pattern with samples that represent all the different soil types or topographic features in the field. A uniform grid (bottom left) can be used if field variability is small or unknown. If there are known sources of variability, such as considerable soil type changes, significant differences in topography, or a history of specific patterns in yield fluctuation across the field, a zone sampling technique (bottom right) is more reasonable, defining zones by those known sources of variation. Sampling patterns can be fine-tuned as new information becomes available over time. The acquisition of a representative sample is perhaps the most important stage in soil testing. Samples are usually collected to the depth of the tillage layer, or to the depth specified by the calibration database used by the soil testing laboratory that will analyze the samples, by hand, using any tool that can produce a uniform sized sample (top-to-bottom), and are usually collected to the depth of the tillage layer, or to the depth specified by the calibration database used by the soil testing laboratory that will analyze the samples. Alternatively, if the number of samples to be collected warrants it, powered hydraulic samplers of various types may be utilized.

SOIL SAMPLING

Reliable soil testing begins with adequate soil sample, regardless of the purpose. The basic ideas of agronomic soil sampling will be outlined in this article, as well as the basic principles of soil sample collection and instructions for developing an effective soil sampling and testing program. Three concepts guide proper soil sample collection:

- ***Organization***: Having a well-organized system for collecting and processing soil samples makes sample collection easier and reduces the risk of human errors like mislabeling or misplacing soil samples.
- ***Consistency***: The quality and dependability of your data will be considerably improved if you collect each sample in a consistent manner throughout years and during a sampling event. This entails collecting samples in the same way each time.
- ***Simplicity***: Simple methods will help you gather samples in a consistent and orderly manner.

Samples and Subsamples

Even over short distances, soils can be extremely varied. Because of this heterogeneity, collecting soil from only one area is frequently insufficient. Collecting so-called composite samples, on the other hand, is preferred. Composite samples are made up of different samples, or subsamples, gathered from several sites and combined to make a single composite sample.

We can reduce the effects of soil variability by averaging soil parameters over greater areas by integrating many subsamples into a single composite sample. Due to concentrated fertilizer treatments (e.g., banded applications) or natural soil variation, composite samples are less sensitive to unusually high or low soil test readings. You'll need the following items and tools before you start collecting soil samples:

- A soil probes
- A clean plastic buckets
- A trowel
- Permanent markers
- Sample bags. Many soil testing laboratories will provide wax-lined sample bags. In lieu of laboratory provided bags, consider using paper bags or zip-top bags.
- Clipboard and paper or field notebook
- GPS-enabled smartphone or handheld GPS unit (optional)

To avoid contamination, it's critical to properly keep composite samples after they've been collected. Most laboratories prefer to prepare samples in their own facilities. As a result, you can frequently send samples directly to the laboratory without having to process them first. Some laboratories demand that samples be sent in certain bags or containers. For details on how to handle and package samples, contact the laboratory you've chosen. If you aren't sending samples to the lab right away, try storing them in the refrigerator or freezer to reduce the likelihood of mold growing in the sample bag. Allow the samples to air dry slightly if the soil is very moist or you don't have access to a refrigerator or freezer. Spread the dirt in a thin layer on a flat surface like a table. To protect the surface from becoming soiled, lay down some paper, such as used newspaper. Excessive heat might harm a sample and change laboratory results, therefore never dry a sample in an oven or microwave.

Sampling Locations and Strategies

Where you gather soil samples is primarily determined by the management technique you use on your farm. There are two sorts of management strategies: whole-field and spatially explicit management. The field is controlled as a single unit under whole-field management. One fertilizer rate is applied consistently across the entire field when fertilizer is applied in a whole-field manner. Whole-field management is simple to set up and requires no special equipment or data processing. The field is divided into smaller portions and each section is managed separately in spatially explicit management. Spatially explicit management can identify parts of the field that require specific fertilizer or liming requirements, as well as produce a map of the field's nutrient and liming demands. Precision agriculture necessitates spatially explicit management. Variable-rate technology can be used in spatially explicit management to change fertilizer and lime treatments so that each zone receives a targeted, zone-specific fertilizer or lime application. In the end, the choice between whole-field and spatially explicit management will be determined by each operation's individual aims and limits. Whole-field management is straightforward to apply, but it runs the danger of over- or under-fertilizing certain portions of the field. Overapplication of fertilizer and lime can result in wasted material and environmental damage, while underapplication might result in low output. On the other hand, geographically explicit management can be more difficult to implement, but the spatial information can be used to reduce the danger of fertilizer and lime overuse and underuse.

Soil sampling for whole-field management: Soil sampling for whole-field management has the purpose of obtaining a sample that is representative of the normal soil in the field. To accomplish this, subsamples are dispersed over broad areas to ensure that the entire field is covered. Collect composite samples in a zig-zag manner to achieve this. Each composite sample should be made up of ten to twenty subsamples that are evenly distributed across a field. At least one composite per 20 acres should be collected.

Soil sampling for spatially explicit management: In spatially explicit management, there are two basic approaches for soil sampling: zone-based sample and grid sampling. Soil samples are taken from predetermined regions in a field using each procedure. You can create a map of soil fertility trends in the field by comparing soil test results with the area of the field where samples were gathered. The smaller the predefined region (i.e., zone or grid cell) in both methods, the more detailed the generated map.

***Zone-based sampling*:** The purpose of zone-based techniques is to gather samples that are representative of the average soil in each zone. Zones are usually created based on specific soil types or yield map patterns. The fact that zones represent areas of homogeneous or regular soil conditions is the most important characteristic of each zone. In order to effectively map soil fertility patterns, the zones typically need to be smaller the more variable the soil is. Zones are typically between 2 and 10 acres in size. Collect a composite sample from each zone for zone sampling. Larger zones necessitate a greater number of samples than smaller zones. Collect two subsamples per acre in a zone as a rule of thumb. A minimum of five subsamples per zone should be collected, regardless of zone size. Within a zone, subsamples should be distributed at random. The true value of zone-based sampling is the ability to generate a map of soil fertility and pH within a field using zone-based data. This map can be used to construct precise, variable-rate fertilizer and lime applications by highlighting areas of exceptionally high or low fertility. Because zone-based samples will eventually be used to create a map, it's critical that samples are labeled with the correct zone and that the position of each sample or zone is documented. This will ensure that the laboratory results from each soil sample are accurately connected with the relevant field site.

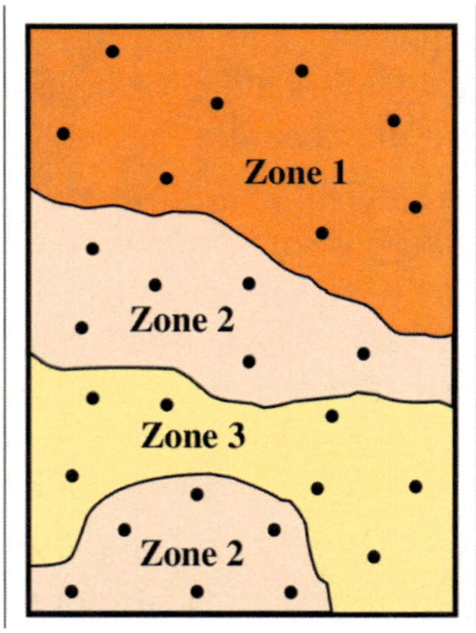

Figure 15: Zone based sampling

Source: Ferguson & Hergert, (2000)

***Grid sampling*:** Many farmers and agriculture service providers use a grid-based approach for soil sampling. Soil samples are gathered at uniformly spaced places in a grid system. Grid sizes typically range from 1 to 5 acres, with smaller grids (2.5 acres or fewer) delivering the best results. Collect at least five subsamples for each one-acre grid cell. Collect 8 to 10 subsamples for grid cells between 2.5 and 5 acres. Grids are appealing because they are simple to use and provide better spatial data than whole-field soil sample. Grid-sampling, on the other hand, does not consider or account for underlying heterogeneity in management zones or soil types. As a result, when compared to zone-based sampling, grid sampling has significant drawbacks. Grid sampling has several drawbacks, the first of which is an increased chance of bias. If the bulk of grid samples fall on a single soil type, for example, the grid sample may underrepresent the underlying variability in the field. Grid sampling can be inefficient as well. Since grids do not target areas of known variation in a field, additional samples are required to cover the whole range of soil types and variability. While grid sampling is preferable than whole-field sampling, it pales in comparison to zone-based sampling, which is more informative.

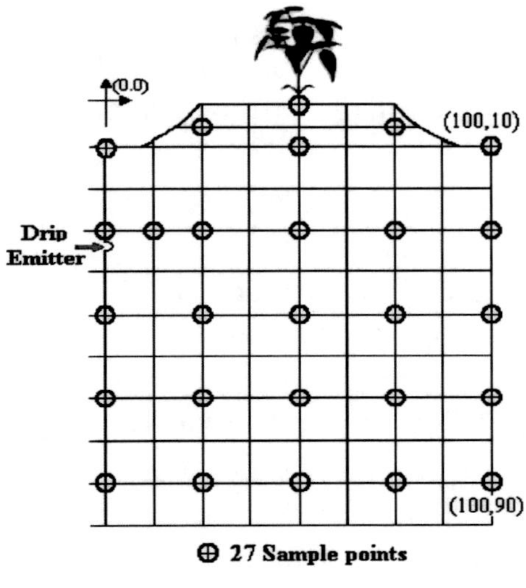

Figure 16: Grid soil sampling method

Source: Bufon, (2010)

Areas to avoid When Sampling

There are some regions you may want to avoid when collecting samples in the field or identifying soil-sampling locations, regardless of your soil sampling approach. These are problem regions in the field where the soil may not be reflective of the rest of the soil in a field or zone due to some disturbance of natural variance. If your soil sampling point is located within or near one of these problem regions, consider moving it to a different location. Consider sampling each issue region as a separate zone if it takes up a big section of a field (more than 5-10 acres).

- End-rows or areas of heavy equipment traffic
- Wet spots or depressions
- Highly eroded areas
- Locations of former farmsteads or animal enclosures
- Locations within 100 ft. of roads
- Areas where lime is stored before application

Sample Timing

When there is no standing crop in the field, it is the most convenient period to take soil samples (i.e. in the spring prior to planting or in the fall after harvest). For a given management program, both sample collection times can be beneficial. Consider the following factors when choosing a sample collection time:

- Prior to applying fertilizers or lime, conduct soil sampling and testing.
- Collect samples early enough for the laboratory to return the results in a timely manner.
- To avoid year-to-year variability, keep sample collection times consistent. For example, if spring sampling is used one year, do not switch to fall sample the next year.

Try to sample fields every four years in most soils. This should be enough information to track changes in soil fertility. This rule does not apply to sandy soils or soils that have been intensively treated (e.g., irrigated fields, silage, or hayfields, etc.). Soil fertility can change quickly in these conditions, so these fields should be sampled every two years. Soil sampling and testing can be quite useful. Monitoring changes in soil fertility, formulating fertilizer recommendations, and improving on-farm nutrient efficiency can all benefit from data from a well-conducted soil-sampling event. Soil sampling utilizing a consistent, well-conducted, and organized approach will yield the most useable and interesting soil test findings, regardless of the soil-sampling program you choose.

TYPES OF SOIL TESTS

Soil testing is analyzing soil samples to evaluate properties such as texture, stability, plant nutrients, fertility, acidity, or alkalinity, as well as toxicities and pollutants. Soil testing results can be useful for a variety of objectives, including agricultural and horticultural production, geotechnical and engineering, and environmental forensics. Soil tests can provide answers to several questions, including:

- Is this soil suitable for growing a lucrative crop, sustaining animals, or revegetation following development or disturbance?
- What fertilizer or ameliorants should I use, and how much of each should I use, to produce the best yields for my crops while avoiding soil contamination?
- If I create foundations or designed soil structures like dams or banks out of this soil material, will they be stable?
- Is this soil hazardous to the environment because of its peculiar properties, such as salinity or the potential for Acid Sulfate Soil?
- Is this soil susceptible to wind or water erosion, and will it result in water contamination or sedimentation downstream?
- Is this soil a health threat because of the presence of toxic chemicals or pollutants like lead, herbicides, or pesticides?

For large-scale production, a thorough research of the raw material is always recommended, as is a full laboratory analysis. However, extensive tests are not required to evaluate a soil's suitability for small-scale production. Simple preliminary tests can be performed on-site to determine the components of a soil sample, including the silt/clay and sand fractions, as well as to study critical characteristics.

Preliminary on-site tests

Smell test: The presence of organic material is indicated by damp soil generating a musty stench.

Color appearance: Organic matter is the dark brown granular humus in the soil. The presence of light brown to black coloration in the soil suggests that it includes at least a small amount of organic matter. However, the color test does not always work. Black cotton soils, for example, are dark brown to black in appearance but contain little organic matter. The presence of iron oxides is indicated by a reddish to dark brown tint. The presence of lime-based chemicals or sand is indicated by the white to yellow coloration. The presence of clay is indicated by a pale brown color.

Shine test: To identify the major component in the sample, a little bit of dry dirt is scraped with the back of a finger nail. If sand or silt is predominant, the soil surface is abrasive to the touch and the soil remains dull. A sample containing clay, on the other hand, sparkles and feels smooth to the touch.

*Thread rolling test***:** This test requires a modest amount of water to be added to a small amount of soil so that it can be easily shaped by hand. The soil sample is then rolled into a thread with the palm of the hand or the fingers on a clean, flat surface. The thread's drop in diameter to around 3 mm suggests the presence of a substantial clay component. The presence of a moderate sand percentage, on the other hand, is indicated by the thread breaking at a wider diameter. This test can also be used to assess a soil's plastic limit.

Hand molding test; The soil sample is moistened and molded into a cube with an edge of about 2.5 cm after stones and foreign bodies larger than about 6 mm diameter have been removed. A large clay fraction is present if a cube is easily constructed.

Laboratory Testing Methods

The research and sampling that go into a laboratory soil test are only as excellent as the research and sampling that goes into it.

*Soil pH Determination***:** In a slurry system, the pH of the soil is normally assessed potentiometrically using an electronic pH meter. Because of the recent profusion of pH meters, an in-depth description of meter operation is not possible; individual devices come with detailed instructions. Several precautions, however, are offered as general guidelines for all labs. To begin, electrodes should be checked and maintained on a regular basis to avoid residue buildup that could interfere with functioning. However, rinsing between each soil sample is rarely essential. Electrodes should be safeguarded in such a way that insertion to the very bottom of a slurry tank does not result in abrasion of the sensing surface, which reduces electrode life and causes false readings. Before use, all meters should be routinely calibrated at two sites. One point of calibration should be at pH 7, while the other should be chosen depending on the lab's typical range of soil pH. A lab that mostly works with acid soils should calibrate over the acid range (for example, the second point at pH 4) while a lab that primarily works with alkaline soils should calibrate across the alkaline range (use second pH buffer at pH 9 or 10). Because flow rates at the liquid junction might affect the accuracy of soil pH readings, reference and/or combination electrodes for measuring soil pH should be carefully chosen. To evaluate the effectiveness of electrodes, labs may choose to employ a set of reference soil samples with a known pH. To avoid changes in soil qualities over time, such samples should be stored and handled under carefully regulated settings. The average reading of multiple meters should be used to calculate the reference soil pH of these samples. Electrodes that fail to yield pH readings consistent with established values when calibrated with clear buffers can be successfully calibrated using reference soils, but only if soil storage and handling are properly controlled. A soil-water slurry is used to determine the pH of the soil. Because the presence of soluble salts in a soil sample might impact pH, some analysts prefer to test pH in a mixture of soil and 0.01 M CaCl2 (Graham, 1959; Schofield and Taylor, 1955). The extra salt in this solution obscures the impact of soluble salt concentration differences in individual samples.

Lime Requirement Determination: The lime requirement processes were created for a variety of soil types. The SMP method was developed for soils with high lime requirements and significant exchangeable Al reserves. The Adams-Evans buffer, on the other hand, was created for coarse-textured soils with low cation exchange capabilities and organic matter levels, resulting in low lime requirements (less than 2 T/A). Although Mehlich calibrated the system for many different soil types, including Histosols, Alfisols, and Inceptisols, the Mehlich lime buffer was created for use on soils in the Southeast US (particularly Ultisols). Hazardous reagents are present in the SMP, Adams-Evans, and Mehlich buffers. All waste materials generated by these methods must be separated and handled separately. Non-regulated reagents have been substituted for the original ingredients in all three buffers. All three alterations were highly correlated with the original buffer and caused no significant changes in analytical results in most cases. Any of the improved methods did not require any calibration modifications. In the United States, the original SMP buffer or its modification (Sikora buffer) is the most widely used approach, with (CT, MA) in the northeast using it. In the mid-Atlantic and coastal plains states, the Adams-Evans is widely used (DE, NJ). ME and PA are now using the Modified Mehlich buffer. In poorly buffered, sandy soils, soils with organic matter contents greater than 10%, or soils with a predominance of kaolinite and Al and Fe oxides in their clay fractions, it is critical to use the correct buffer method for lime requirement, as the SMP has been shown in several studies to be inaccurate. There are several outstanding, thorough sources.

Measurement of Macro and Micronutrients: By substituting the cations on the soil's exchange sites with a counter ion such as Na+ (Morgan) or NH4+, soil tests for cations (Ca, Mg, K) typically estimate the quantity of water-soluble and exchangeable forms (Modified Morgan and Mehlich

3). Because all these extractants are acidic, some non-exchangeable K will be extracted. Furthermore, ammonium-based extractants such as the Modified Morgan and Mehlich 3 solutions extract more K from some soils than the Nabased Morgan extractant. Ca and Mg will be extracted in equal levels in all these soil tests. Because all three are buffered acid extracts, they will also extract non-exchangeable Ca and Mg from soils containing carbonate, either naturally or as a result of recent lime application. The macro and micronutrients in Mehlich 3, Morgan, and modified Morgan soil extracts can be measured using a variety of methods. Phosphorus can be determined using inductively coupled plasma (ICP) or colorimetric methods such as ammonium molybdate or ammonium vanadate; K and Ca can be determined using atomic absorption, atomic emission, or ICP; Mg, Cu, Mn, and Zn can be determined using atomic absorption or ICP; and B can be determined using colorimetric methods or ICP. Although P measured by ICP may be higher than when measured calorimetrically for many soils, the results from the various techniques used to assess macro and micronutrients in soil extracts should, in general, be comparable. Because of this, soil test calibrations should be based on the analytical method that is used most frequently. Although some laboratories employ a colorimetric method for phosphorus, most laboratories use ICPs to measure macro and micronutrients in soil extracts due to their better efficiency.

***Measuring Cation Exchange Capacity*:** The cation exchange capacity (CEC) of a soil is a measurement of the number of negatively charged sites on soil surfaces that can electrostatically retain positively charged ions (cations) like calcium (Ca_2^+), magnesium (Mg_2^+), and potassium (K^+). Because cations maintained electrostatically are easily exchangeable with cations in the soil solution, a soil with a higher CEC can sustain higher levels of Ca_2^+, Mg_2^+, and K^+ than a soil with a lower CEC. Because acid cations such as hydrogen (H+) and aluminum (Al3+) can occupy the CEC, a soil with a greater CEC may not always be more fruitful. CEC, when used in conjunction with other soil fertility indicators, is a good predictor of soil quality and production. Clay minerals and organic matter (OM) surfaces have the most cation exchange sites. When soil OM is near-neutral pH, it develops a higher CEC than when it is acidic (pH-dependent CEC). As a result, adding organic matter to a soil will almost certainly raise its CEC over time. The CEC of a soil, on the other hand, might decline with time due to natural or fertilizer-induced acidity and/or OM decomposition. Soil CEC is usually stated in one of two numerically comparable units: meq/100 g (milliequivalents of charge per 100 g of dry soil) or cmolc/kg (milliequivalents of charge per kilogram of dry soil) (centimoles of charge per kilogram of dry soil). Because of the various methodologies for estimating CEC, it is critical to understand the data's intended application. The CEC of a soil is frequently evaluated at a standard pH value for soil classification reasons. The ammonium acetate method, which is buffered at pH 7, and the barium chloride-triethanolamine method, which is buffered at pH 8.2, are both examples of buffered methods. Such CEC measurements, especially in acidic soils with pH-dependent CEC, might produce values that differ significantly from the CEC of the soil at its field pH (effective CEC or CECe). If you need a pH-buffered CEC measurement (for regulatory or soil categorization purposes, for example), ammonium acetate buffered at pH 7 is the way to go. The CEC as assessed by this approach varies from less than 3cmolc/kg for sandy soils poor in OM to greater than 25cmolc/kg for soils high in particular types of clay minerals or OM in the Northeastern United States. The $BaCl_2^-$ compulsive exchange approach is recommended for a reliable measurement of a soil's CEC under field settings. Under field conditions, this approach estimates the soil's capacity to retain nonacid cations (CECb). Exchangeable Al produced by extraction with 1 M KCl must be added to the CECe estimate. CECe is not commonly measured directly in soil testing laboratories. Rather, the CEC is frequently estimated (calculated, CECsum) using the amounts of Ca_2^+, Mg_2^+, and K^+ extracted in a conventional agronomic soil test (e.g., Mehlich 3 or (modified) Morgan's extraction solution and procedures). Although sodium is not commonly found in substantial concentrations in soils in the Northeast, if it is, extractable Na+ should be included in the calculation. If the soil pH is less than 6, exchangeable acidity is added to the sum of extractable Ca_2^+, Mg_2^+, and K+, indicating the presence of considerable amounts of exchangeable Al3+. Exchangeable acidity is frequently calculated using a regression equation including a buffer pH measurement and 1 M KCl exchangeable acidity (an actual measure of exchangeable Al_3^+ unless the soil is highly acidic or heavy in OM).

Soil Organic Matter Tests: Soil organic matter is widely known for its role in delivering nutrients, contributing to cation exchange capacity, and enhancing soil structure. The amount of organic matter in the soil is utilized in some states to adjust N, S, herbicide, and/or lime recommendations. Organic matter analysis has re-sparked interest due to the importance of soil organic matter in herbicide recommendations. The organic matter content of the soil can also be used to build management strategies for municipal sewage sludge and other wastes that will be applied to the land. Organic matter is commonly determined using one of two methods:

Weight loss on removal of the organic matter from the mineral fraction by:

- Oxidation with H_2O_2
- Ignition
- Ignition after decomposition of silicates with HF

Determination of some constituent that is found in a relatively constant percentage of soil organic matter such as:

- Nitrogen
- Carbon

Volatilization of substances other than organic materials (H_2O, structural OH, CO_2 from carbonates) and incomplete oxidation of carbonaceous materials can induce mistakes in weight loss calculations. Furthermore, these approaches are typically time-consuming. Because of the safety and disposal difficulties with chromic acid, there has been a recent interest in weight loss strategies. Because of the considerable variance in nitrogen content of organic materials from diverse sources, estimating organic matter by determining total nitrogen is not generally employed. Carbon determinations, on the other hand, are often employed for this estimation, with the carbon being measured by:

- Dry combustion and measurement of CO_2 evolved (after removal of carbonates)
- Chromic acid oxidation and measurement of CO_2 evolved (after carbonate removal)
- Chromic acid oxidation to measure easily oxidized material (external heat applied).
- Chromic acid oxidation to measure easily oxidized material (spontaneous heating).

The dry combustion method determines total carbon, while chromic acid methods only determine easily oxidizable carbon. (Chromic acid does not oxidize the carbon in graphite and coal.) Combustion methods that directly detect CO_2 have evolved, but they require specialized equipment and are not well suited to the rapid examination of many samples unless expensive automated and computerized carbon analyzers are utilized. As a result, soil testing laboratories frequently use procedures that rely on chromic acid oxidation to assess easily oxidizable material. The primary difference between these procedures (c and d) is the source and amount of heat needed to fuel the reaction. Technique (c) makes use of an external source of heat, allowing for a higher temperature than method (d), which gets its heat from the heat of dilution of concentrated H_2SO_4. As a result, technique (c) is faster, and organic matter oxidation is more complete, but the circumstances must be carefully managed to produce repeatable results.

A sample list of laboratory soil tests found in the northeastern United States is shown below. Read more about the individual laboratory soil testing processes in the references.

Table 22: Soil Testing Methods

Analyte	Method	Reference
pH	Water	Eckert, D., and J.T. Sims. 2011. Recommended Soil pH and Lime Requirement Tests. p. 19-25. *In* J.T. Sims and A. Wolf (eds.) Recommended Soil Testing Procedures for the Northeastern United States. Northeast Regional Bulletin #493. 3rd edition. Agricultural Experiment Station, University of Delaware, Newark, DE.
Lime requirement	Mehlich buffer	Mehlich, A. 1976. New buffer pH method for rapid estimation of exchangeable acidity and lime requirement of soils. Commun. Soil Sci. Plant Analysis. 7, 637-652.
Extractable P, K, Ca and Mg	Mehlich 3 (ICP)	Wolf, A.M., and D.B. Beegle. 2011. Recommended soil tests for macronutrients. p. 39-47. *In* J.T. Sims and A. Wolf (eds.) Recommended Soil Testing Procedures for the Northeastern United States. Northeast Regional Bulletin #493. 3rd edition. Agricultural Experiment Station, University of Delaware, Newark, DE.
Cation Exchange Capacity (CEC)	Summation	Ross, D., and Q. Ketterings. 2011. Recommended soil tests for determining soil cation exchange capacity. p. 75-86. *In* J.T. Sims and A. Wolf (eds.) Recommended Soil Testing Procedures for the Northeastern United States. Northeast Regional Bulletin #493. 3rd edition. Agricultural Experiment Station, University of Delaware, Newark, DE.

Organic matter	Loss on Ignition	Schulte, E.E. and B. Hoskins. 2011. Recommended Soil Organic Matter Tests. p. 63-74. *In* J.T. Sims and A. Wolf (eds.) Recommended Soil Testing Procedures for the Northeastern United States. Northeast Regional Bulletin #493. 3rd edition. Agricultural Experiment Station, University of Delaware, Newark, DE.
Total C	Combustion	Nelson, D.W. and L.E. Sommers. 1996. Total Carbon, Organic Carbon, and Organic Matter. p 961-1010. *In* D.L. Sparks (ed). Methods of Soil Analysis, Part 3. Chemical Methods. Soil Science Society of America Book Series Number 5. American Society of Agronomy, Madison, WI. Pella, E. 1990. Elemental organic analysis. Part 1. Am. Lab 22: 116-125
Nitrate N	Specific Ion Electrode	Griffin, G., W. Jokela, D. Ross, D. Pettrinelli, T. Morris, and A. Wolf. 2011. Recommended Soil Nitrate-N Tests. p. 27-38. *In* J.T. Sims and A. Wolf (eds.) Recommended Soil Testing Procedures for the Northeastern United States. Northeast Regional Bulletin #493. 3rd edition. Agricultural Experiment Station, University of Delaware, Newark, DE.
Ammonium N	Specific Ion Electrode	Mulvaney, R.L. 1996. Nitrogen-Inorganic Forms. p. 1123-1200. *In* D.L. Sparks (ed). Methods of Soil Analysis, Part 3. Chemical Methods. Soil Science Society of America Book Series Number 5. American Society of Agronomy, Madison, WI.
Total N	Combustion	Bremner, J.M. 1996. Nitrogen-Total. p. 1085-1121. *In* D.L. Sparks (ed). Methods of Soil Analysis, Part 3. Chemical Methods. Soil Science Society of America Book Series Number 5. American Society of Agronomy, Madison, WI. Pella, E. 1990. Elemental organic analysis. Part 1. Am. Lab 22: 116-125

Soluble Salts	Electrical Conductivity (1:2)	Gartley, K. 2011. Recommended Soluble Salts Tests. p. 87-94. *In* J.T. Sims and A. Wolf (eds.) Recommended Soil Testing Procedures for the Northeastern United States. Northeast Regional Bulletin #493. 3rd edition. Agricultural Experiment Station, University of Delaware, Newark, DE.
Total Sorbed Cu, Zn, Pb, Ni, Cd, Cr, Mo	EPA Method 3050B/3051 + 6010	USEPA. 1986. Test Methods for Evaluating Solid Waste. Volume IA: 3rd Edition. EPA/SW-846. National Technical Information Service. Springfield, Va.
Particle Size Analysis	Hydrometer Method	Gee, G.W. and J.W. Bauder. 1986. Particle size analysis. p. 383-411. *In* A. Klute (ed.) Methods of Soil Analysis. Part 1. Physical and Mineralogical Methods. Agronomy Monograph #9 (2nd Edition). Amer. Soc. Agron. Madison, WI.
As	EPA Method 3050B/3051 + 6010	USEPA. 1986. Test Methods for Evaluating Solid Waste. Volume IA: 3rd Edition. EPA/SW-846. National Technical Information Service. Springfield, Va.
Se	EPA Method 3050B/3051 + 6010	USEPA. 1986. Test Methods for Evaluating Solid Waste. Volume IA: 3rd Edition. EPA/SW-846. National Technical Information Service. Springfield, Va.
Hg	EPA Method 7473	USEPA. 1986. Test Methods for Evaluating Solid Waste. Volume IA: 3rd Edition. EPA/SW-846. National Technical Information Service. Springfield, Va.
Calcium carbonate equivalency (CCE)	ASTM Method C 25	Standard Test Methods for Chemical Analysis of Limestone, Quicklime, and Hydrated Lime, ASTM International, 100 Barr Harbor Drive, PO Box C700, West Conshohocken, PA

Source: https://agsci.psu.edu/aasl/soil-testing/methods

POLICY AND FERTILIZER USE

Fertilizer use necessitates prudence. Improper and excessive use can have a negative impact on the environment. Nitrate levels in drinking water can be detrimental to human health, especially in newborns under the age of six months. Nitrate leaching has been found to be strongly linked to nitrogen inputs that exceed the agronomic maximum. Eutrophication is produced by nitrate and phosphate deposits in lakes, ponds, and other bodies of water, resulting in excessive algae growth, oxygen deprivation, and fish mortality. Plants absorb cadmium from phosphate fertilizer made from cadmium-rich phosphate rocks, but it's unclear how cadmium gets from phosphate fertilizer to food crops and then to humans. Excessive levels of nitrate and phosphate in soil and water bodies are to blame for these environmental issues. Organic fertilizer, animal dung, and industrial and urban wastes are all suitable sources of these nutrients, in addition to mineral fertilizer. Animal dung has been related to high amounts of nitrate in water in numerous European countries. High nitrate levels have also been connected to sewage disposal, septic tank drainage, and industrial wastes in underdeveloped countries. As a result, the issue is one of integrated nutrient management in agricultural areas, with the goal of keeping total nutrient supply below the agronomic maximum. Another management method for reducing pollution and maintaining soil productivity is to prevent erosion and runoff. Previously, the focus was on increasing fertilizer use; today, the focus must change to educating farmers on how to use organic, inorganic, and biological fertilizer to their full potential. The widespread use of nitrogen-based fertilizers in developing countries has resulted in nutritional imbalances in soils. Nutrient balance should be enhanced by boosting the use of phosphate and potash fertilizers to improve nitrogen fertilizer efficiency and avoid negative environmental effects. Furthermore, 50 to 60% of fertilizer nutrients given to the soil are lost to the atmosphere. Nutrient losses can be decreased by using fertilizer products at the right time, applying them correctly, and placing them in the right places, as well as minimizing soil erosion and water runoff. Reduced nutrient losses will benefit both the economy and the environment. Even though fertilizer use has expanded dramatically in recent decades, particularly in East Asia, many developing countries' fertilizers use is still insufficient to cause environmental damage. To avoid potential damage, procedures to monitor environmental impact are essential locations, such as Java in Indonesia, Punjab in India, and the Delta region in Egypt, where application rates are high. Nitrate leaching does not occur from flooded paddy fields, which account for 60 to 70% of nitrogen fertilizer use in East

Asian countries due to low percolation rates, but nitrate contamination of rivers and lakes produced by nitrogen runoff from such areas requires special care. Nitrogen applications should be carefully handled in irrigated and high-rainfall locations with light-textured soils and vegetable crops, as these conditions encourage nitrate leaching. Nitrate levels in drinking water should not exceed 50 milligrams per liter of water, according to the World Health Organization; where levels are greater, location-specific measures should be implemented. Fertilizer production, particularly nitrogen production, is energy intensive, according to opponents, and scarce energy resources should not be wasted on fertilizer manufacture. Fertilizer, on the other hand, contributes for a minor percentage of global energy consumption (2% in 1990). An additional 218 kilograms of grain—enough to supply the minimum calorie intake for one person per year—could be produced for every 1 million Btus of energy used in the fertilizer sector (equal to the energy needed in driving a car from Washington, D.C., to New York City). Natural gas prices in various energy-exporting countries averaged around a dollar per million Btus in 1990. As a result, transforming energy into food security through the application of fertilizers is a cost-effective and humanitarian way of utilizing the world's energy resources.

Because of underdeveloped commercial markets, a lack of private investment, and worries about food security, fertilizer production, import, and marketing has traditionally been a government activity in most developing nations. Protection, subsidies, and price controls have defined the fertilizer industry. Although this has aided the development of fertilizer markets, resource inefficiencies and unsustainable fiscal burdens have forced a shift. Markets should be liberalized, subsidies withdrawn, and the sector privatized to boost efficiency in nations where fertilizer consumption is high and the private sector and financial markets are highly developed. To make fertilizer markets more competitive, policy and organizational reforms are required, including gradual liberalization of trade restrictions to allow the private sector to participate in the worldwide market. Because the simultaneous introduction of various policy reforms, such as devaluation, subsidy removal, and closure of public sector enterprises, can result in a drastic reduction in fertilizer use, as happened in Ghana, Poland, Russia, and Zambia, proper sequencing and phasing of these reforms may pose a challenge. To fully comprehend the dynamics of policy and organizational reforms, more research is required. The elimination of fertilizer subsidies and crop price support programs will lead to the convergence of economic and environmental goals by promoting resource use efficiency, reducing fiscal deficits, and minimizing environmental damage in areas where fertilizer subsidies or crop price support programs have encouraged excessive fertilizer use. Although energy consumption is required for fertilizer manufacturing, the efficiency of energy consumption must be enhanced. Fertilizer plants in many emerging and reforming countries utilize 37 to 61 million Btus per ton of ammonia output, compared to roughly 30 million Btus in modern plants. The proper operation and maintenance of facilities, as well as the replacement of aging technologies with new energy-efficient technologies, are all necessary. Over the next few years, the policy challenge will be to control environmental issues in high-use areas without sacrificing fertilizer's productivity benefits. The problem in low-usage areas is to boost fertilizer use in an environmentally sustainable way. The following policies should be implemented to ensure a favorable and stable policy environment for fertilizer use and supply:

- ***Macroeconomic stability*:** Macroeconomic stability, particularly exchange rate stability, is critical for fostering fertilizer consumption and supply growth. By raising costs and lowering investor confidence, rapid depreciation of the domestic currency limits fertilizer use and supply. Because many developing countries lack sufficient fertilizer supply, proper and timely foreign exchange allocation for fertilizer imports should be a top priority.
- ***Pricing policy:*** The pricing policy should be managed in such a way that it provides appropriate incentives for small farmers to utilize fertilizer. Any price inconsistencies should be removed. Although fertilizer subsidies should be phased down gradually, there is a case to be made for a temporary subsidy in landlocked and food-deficit countries with skewed markets, limited infrastructure, positive environmental externalities, and widespread poverty.
- ***Efficient and appropriate organizations*:** To ensure that fertilizer reaches the farm on time, in suitable quantities, and at a low cost, efficient and appropriate organizations should be established. The major duty for fertilizer marketing and distribution should fall to the private sector. For the fertilizer markets to function efficiently, the government should design and implement suitable regulatory and quality control procedures. The government may take the lead in building markets and supporting infrastructure in places where markets are undeveloped.

- ***Funding***: Fertilizer use is severely limited due to a lack of finances available to farmers to purchase fertilizers. Because the private sector is increasingly involved in fertilizer marketing and distribution, fertilizer dealers must have access to financial resources as well. Both farmers and fertilizer dealers should make every effort to secure appropriate finances at acceptable interest rates.
- ***Investment***: The government should create a market-friendly environment to attract capital investment in fertilizer production and imports. Fertilizer self-sufficiency should not be a prized objective in and of itself. Joint ventures between wealthy countries with advanced technologies and developing countries with abundant natural resources should be given more attention.
- ***R&D:*** Farmers should be provided with adequate research, extension, and instructional support. Soil testing and new technology that allow for precise fertilizer application should be supported.
- ***Monitoring***: Environmental monitoring tools should be established, as well as corrective actions. To reduce negative environmental effects, the adoption of appropriate practices and technologies should be encouraged. Fertilizer use will continue to be an important part of future farming techniques.

Food security, as well as the preservation of the natural resource base Fertilizer use, on the other hand, should be addressed differently in the future in order to fulfill that purpose. So that the greater aims of food security, agricultural expansion, and environmental protection are not jeopardized, the focus should be on growth with management rather than growth per se.

References

1. Bridges, E. M. (1993). Soil horizon designations; past use and future prospects. Catena, 20(4), 363-373.
2. Khaing, T., Win, K. H., & Khaing, Y. K. (2019). Analysis of Important Parameter in Soil Samples from Selected Areas. Technological University (Myitkyina) 1st University Research Journal of Science, Engineering and Engineering Education, (URJSEE) Volume 01, 156.
3. Kellogg, C. E. (1993). Soil survey division staff: soil survey manual. United States Department of Agriculture, Washington.
4. Brady, N. C., Weil, R. R., & Weil, R. R. (2008). The nature and properties of soils (Vol. 13, pp. 662-710). Upper Saddle River, NJ: Prentice Hall.
5. Edwards, Clive A., and P.J. Bohlen. (1996). Biology and Ecology of Earthworms. Chapman and Hall, New York. 426 p.
6. Jackson, M. L., & Sherman, G. D. (1953). Chemical weathering of minerals in soils. Advances in agronomy, 5, 219-318.
7. Hallberg, G. R. (1984). The US system of soil taxonomy: From the outside looking in. Soil Taxonomy—Achievements and Challenges, 14, 45-59.
8. Agriculture Organization of the United Nations. Soil Resources, Management, Conservation Service, Agriculture Organization of the United Nations. Land, & Water Development Division. (1996). Agro-ecological zoning: Guidelines (No. 73). Food & Agriculture Org.

9. Sanchez, P. A., & Jama, B. A. (2002). Soil fertility replenishment takes off in East and Southern Africa. Integrated plant nutrient management in sub-Saharan Africa: From concept to practice, CABI, Wallingford, UK, 23-46.
10. Lal, R., Iivari, T., & Kimble, J. M. (2003). Soil degradation in the United States: extent, severity, and trends. CRC Press.
11. Dent, D. L. (1986). Acid sulphate soils: a baseline for research and development (No. 39). ILRI.
12. Oldeman, L. R. (1992). Global extent of soil degradation. In Bi-Annual Report 1991-1992/ISRIC (pp. 19-36). ISRIC.
13. Tan, Z. X., Lal, R., & Wiebe, K. D. (2005). Global soil nutrient depletion and yield reduction. Journal of sustainable agriculture, 26(1), 123-146.
14. Henao, J., & Baanante, C. A. (1999). Estimating rates of nutrient depletion in soils of agricultural lands of Africa. Muscle Shoals: International Fertilizer Development Center.
15. Smaling, E. M. A., & Fresco, L. O. (1993). A decision-support model for monitoring nutrient balances under agricultural land use (NUTMON). Geoderma, 60(1-4), 235-256.
16. Sheldrick, W. F., Syers, J. K., & Lingard, J. (2002). A conceptual model for conducting nutrient audits at national, regional, and global scales. Nutrient cycling in Agroecosystems, 62(1), 61-72.
17. FAO, UNDP, UNEP, World Bank (1997) Land quality indicators and their use in sustainable agriculture and rural development. FAO, Rome
18. Trenkel, M. E. (2010). Slow-and controlled-release and stabilized fertilizers: an option for enhancing nutrient use efficiency in agriculture. IFA, International fertilizer industry association.
19. Dobermann, A., Santa Cruz, P. C., & Cassman, K. G. (1995, February). Potassium balances and soil potassium supplying power in intensive irrigated rice ecosystems. In 24th Colloquium of IPI (pp. 199-229).
20. Mutert, E. W. (1996). Plant nutrient balances in asia and pacific region: facts and consequences for agricultural production. In APO-FFTC seminar on appropriate use of fertilizers (pp. 73-112).
21. Havlin, J. L., Beaton, J. D., Tisdale, S. L., & Nelson, W. L. (1999). Soil Fertility and Fertilizers Prentice Hall. New Jersely, 345-355.
22. Robertson, F. N. (1989). Arsenic in ground-water under oxidizing conditions, south-west United States. Environmental geochemistry and Health, 11(3-4), 171-185.

23. Osman, K. T. (2014). Chemical soil degradation. In Soil degradation, conservation and remediation (pp. 125-148). Springer, Dordrecht.
24. Pieri, C. (1989). Fertilité des terres de savanes. Bilan de trente ans de recherche et de développement agricoles au sud du Sahara. CIRAD-IRAT.
25. Heard, J., Cavers, C., & Adrian, G. (2006). Up in smoke-nutrient loss with straw burning. Better Crops, 90(3), 10-11.
26. Biederbeck, V. O., Campbell, C. A., Bowren, K. E., Schnitzer, M., & McIver, R. N. (1980). Effect of burning cereal straw on soil properties and grain yields in Saskatchewan. Soil Science Society of America Journal, 44(1), 103-111.
27. Gascho, G. J., Davis, J. G., Hook, J. E., Wauchope, R. D., Dowler, C. C., Johnson, A. W., ... & Sumner, H. R. (1998). Nitrate-nitrogen, soluble, and bioavailable phosphorus runoff from simulated rainfall after fertilizer application. Soil Science Society of America Journal, 62(6), 1711-1718.
28. Mello, E. L. (2002). Erosão hídrica em diferentes sistemas de manejo do solo sob chuva simulada. Lages, SC, UDESC.
29. Burwell, R. E., Timmons, D. R., & Holt, R. F. (1975). Nutrient transport in surface runoff as influenced by soil cover and seasonal periods. Soil Science Society of America Journal, 39(3), 523-528.
30. Johnson, H. P., Baker, J. L., Shrader, W. D., & Laflen, J. M. (1979). Tillage system effects on sediment and nutrients in runoff from small watersheds. Transactions of the ASAE, 22(5), 1110-1114.
31. Schick, J., Bertol, I., Batistela, O., & Balbinot Júnior, A. A. (2000). Erosão hídrica em Cambissolo Húmico alumínico submetido a diferentes sistemas de preparo e cultivo do solo: I. Perdas de solo e água. Revista Brasileira de Ciência do Solo, 24, 427-436.
32. Gafur, A., Borggaard, O. K., Jensen, J. R., & Petersen, L. (2000). Changes in soil nutrient content under shifting cultivation in the Chittagong Hill Tracts of Bangladesh. Geografisk Tidsskrift-Danish Journal of Geography, 100(1), 37-46.
33. Upjohn B, Fenton G, Conyers M (2005) Soil acidity and liming Agfact AC.19, 3rd edn. NSW Department of Primary Industries. Orange, NSW.
34. Barak, P., Jobe, B. O., Krueger, A. R., Peterson, L. A., & Laird, D. A. (1997). Effects of long-term soil acidification due to nitrogen fertilizer inputs in Wisconsin. Plant and soil, 197(1), 61-69.
35. Glendinning, J. S. (2000). Australian soil fertility manual. CSIRO Pub..
36. Watmough, S. A., & Dillon, P. J. (2003). Calcium losses from a forested catchment in south-central Ontario, Canada. Environmental science & technology, 37(14), 3085-3089.
37. Fenn, M. E., Huntington, T. G., McLaughlin, S. B., Eager, C., Gomez, A., & Cook, R. B. (2006). Status of soil acidification in North America. Journal of Forest Science 52: 3-13.

38. Matzner, E., & Davis, M. (1996). Chemical soil conditions in pristine Nothofagus forests of New Zealand as compared to German forests. Plant and Soil, 186(2), 285-291.

39. Slattery, W. J., Conyers, M. K., & Aitken, R. L. (1999). Soil pH, aluminium, manganese and lime requirement. Soil analysis: an interpretation manual, 103-128.

40. Slattery, M. L., Benson, J., Curtin, K., Ma, K. N., Schaeffer, D., & Potter, J. D. (2000). Carotenoids and colon cancer. The American journal of clinical nutrition, 71(2), 575-582.

41. Yokota, S., & Ojima, K. (1995). Physiological response of root tip of alfalfa to low pH and aluminium stress in water culture. In Plant-Soil Interactions at Low pH: Principles and Management (pp. 329-331). Springer, Dordrecht.

42. Kennedy, C. W., & Jones, J. E. (1991). Evaluating quantitative screening methods for manganese toxicity in cotton genotypes. Journal of plant nutrition, 14(12), 1331-1339.

43. Mortley, D. G. (1993). Manganese toxicity and tolerance in sweetpotato. HortScience, 28(8), 812-813.

44. Mgema, W. G., & Clark, R. B. (1995). Sorghum genotypic differences in tolerance to excess manganese. Journal of plant nutrition, 18(5), 983-993.

45. Taylor, G. J., Stadt, K. J., & Dale, M. R. (1991). Modelling the phytotoxicity of aluminum, cadmium, copper, manganese, nickel, and zinc using the Weibull frequency distribution. Canadian Journal of Botany, 69(2), 359-367.

46. Jett, J. W. (2005). Plant pH preferences. West Virginia Univ. Ext. Serv., Morgantown, WV.

47. Slattery, J. F., Coventry, D. R., & Slattery, W. J. (2001). Rhizobial ecology as affected by the soil environment. Australian Journal of Experimental Agriculture, 41(3), 289-298.

48. Siqueira, J. O., & Moreira, F. M. S. (1997). Microbial populations and activities in highly-weathered acidic soils: highlights of the Brazilian research. MONIZ, AC; FURLANI, AMC; SCHAFFERT, RE; FAGERIA, NK, 139-156.

49. Mele, P. M., & Carter, M. R. (1999). Impact of crop management factors in conservation tillage farming on earthworm density, age structure and species abundance in south-eastern Australia. Soil and Tillage Research, 50(1), 1-10.

50. Lavelle, P., & Spain, A. V. (2001). Soil ecology. Springer Science & Business Media.

51. Ghassemi, F., Jakeman, A. J., & Nix, H. A. (1995). Salinisation of land and water resources: human causes, extent, management and case studies. CAB international.

52. Ahuja, L. R. (1983). Modeling Infiltration into Crusted Soils by the Green—Ampt Approach. Soil Science Society of America Journal, 47(3), 412-418.

53. McIntyre, D. S. (1958). Permeability measurements of soil crusts formed by raindrop impact. Soil science, 85(4), 185-189.

54. Agassi, M., Shainberg, I., & Morin, J. (1981). Effect of electrolyte concentration and soil sodicity on infiltration rate and crust formation. Soil Science Society of America Journal, 45(5), 848-851.

55. Assouline, S. (2006). Modeling the relationship between soil bulk density and the water retention curve. Vadose Zone Journal, 5(2), 554-563.

56. Lynch, J. M., & Bragg, E. (1985). Microorganisms and soil aggregate stability. In Advances in soil science (pp. 133-171). Springer, New York, NY.

57. Heil, J. W., Juo, A. S., & McInnes, K. J. (1997). Soil properties influencing surface sealing of some sandy soils in the Sahel. Soil Science, 162(7), 459-469.

58. Betzalel, I., Morin, J., Benyamini, Y., Agassi, M., & Shainberg, I. (1995). Water drop energy and soil seal properties. Soil Science, 159(1), 13-22.

59. Zejun, T., Tingwu, L., Qingwen, Z., & Jun, Z. (2002, May). The sealing process and crust formation at soil surface under the impacts of raindrops and polyacrylamide. In Proc. 12th Conf. ISCO, May (pp. 26-31).

60. Wang, Z. L., Shao, M. A., & Chang, Q. R. (1998). Effects of rainfall factors on soil erosion in Loess Plateau. Acta Universitatis Agriculturalis Boreali-occidentalis, 26(4), 101-105.

61. Remley, P. A., & Bradford, J. M. (1989). Relationship of soil crust morphology to inter-rill erosion parameters. Soil Science Society of America Journal, 53(4), 1215-1221.

62. Northcote, K. H. (1971). Factual key for the recognition of Australian soils.

63. Lal, R. (Ed.). (2006). Encyclopedia of soil science (No. 11). CRC Press.
64. Collis-George, N., & Greene, R. S. B. (1979). The effect of aggregate size on the infiltration behaviour of a slaking soil and its relevance to ponded irrigation. Soil Research, 17(1), 65-73.
65. Mullins, C. E., Young, I. M., Bengough, A. G., & Ley, G. J. (1987). Hard-setting soils. Soil use and management, 3(2), 79-83.
66. Kemper, W. D., Rosenau, R. C., & Dexter, A. R. (1987). Cohesion development in disrupted soils as affected by clay and organic matter content and temperature. Soil Science Society of America Journal, 51(4), 860-867.
67. Fabiola, N., Giarola, B., da Silva, A. P., Imhoff, S., & Dexter, A. R. (2003). Contribution of natural soil compaction on hardsetting behavior. Geoderma, 113(1-2), 95-108.
68. Chartres, C. J., Kirby, J. M., & Raupach, M. (1990). Poorly ordered silica and aluminosilicates as temporary cementing agents in hard-setting soils. Soil Science Society of America Journal, 54(4), 1060-1067.
69. Oster, J. D., Shainberg, I., & Wood, J. D. (1980). Flocculation value and gel structure of sodium/calcium montmorillonite and illite suspensions. Soil Science Society of America Journal, 44(5), 955-959.
70. Levy, G. J., Levin, J., & Shainberg, I. (1994). Seal formation and interrill soil erosion. Soil Science Society of America Journal, 58(1), 203-209.
71. Le Bissonnais, Y., Fox, D., & Bresson, L. M. (1998). Incorporating crusting processes in erosion models. In Modelling soil erosion by water (pp. 237-246). Springer, Berlin, Heidelberg.
72. Commission of the European Communities (2006). Proposal for a Directive of the European Parliament and of the Council establishing a framework for the protection of soil and amending Directive 2004/35/EC. COM (2006) 232 final http://ec.europa.eu/environment
73. Soane, B. D., & Van Ouwerkerk, C. (1995). Implications of soil compaction in crop production for the quality of the environment. Soil and Tillage Research, 35(1-2), 5-22.
74. Eriksson, J., Håkansson, I., & Danfors, B. (1974). Jordpackning-markstruktur-gröda [The effect of soil compaction on soil structure and crop yields]. In Rep. 354 (pp. 1-82). Swedish Institute of Agricultural Engineering.

75. Kayombo, B., & Lal, R. (1994). Responses of tropical crops to soil compaction. In Developments in Agricultural Engineering (Vol. 11, pp. 287-316). Elsevier.

76. Gill, W. R. (1971). Economic assessment of soil compaction. Compaction of agricultural soils.

77. Van den Akker, J. J. H., & Schjønning, P. (2004). Subsoil compaction and ways to prevent it. Managing soil quality: Challenges in modern agriculture, 163-184.

78. Soane, B. D., Dickson, J. W., & Campbell, D. J. (1982). Compaction by agricultural vehicles: A review III. Incidence and control of compaction in crop production. Soil and Tillage Research, 2(1), 3-36.

79. Lipiec, J., & Stępniewski, W. (1995). Effects of soil compaction and tillage systems on uptake and losses of nutrients. Soil and Tillage Research, 35(1-2), 37-52.

80. Alakukku, L., Weisskopf, P., Chamen, W. C. T., Tijink, F. G. J., Van Der Linden, J. P., Pires, S., ... & Spoor, G. (2003). Prevention strategies for field traffic-induced subsoil compaction: a review: Part 1. Machine/soil interactions. Soil and tillage research, 73(1-2), 145-160.

81. Eriksson, J. (1982). Markpackning och rotmiljö (Soil compaction and root environment). In Rep. 126 (p. 138). Swedish University of Agricultural Sciences, Division of Agricultural Hydrotechnics Uppsala.

82. Ehlers, W. (1982). Die Bedeutung des Bodengefüges für das Pflanzenwachstum bei moderner Landbewirtschaftung. Mitteilungen Deutsche Bodenkundliche Gesellschaft, 34, 115-128.

83. Arvidsson, J. (1998). Influence of soil texture and organic matter content on bulk density, air content, compression index and crop yield in field and laboratory compression experiments. Soil and Tillage Research, 49(1-2), 159-170.

84. Blackwell, P. S., Graham, J. P., Armstrong, J. V., Ward, M. A., Howse, K. R., Dawson, C. J., & Butler, A. R. (1986). Compaction of a silt loam soil by wheeled agricultural vehicles. I. Effects upon soil conditions. Soil and Tillage Research, 7(1-2), 97-116.

85. Pietola, L., Horn, R., & Yli-Halla, M. (2005). Effects of trampling by cattle on the hydraulic and mechanical properties of soil. Soil and tillage research, 82(1), 99-108.

86. Alakukku, L., Weisskopf, P., Chamen, W. C. T., Tijink, F. G. J., Van Der Linden, J. P., Pires, S., ... & Spoor, G. (2003). Prevention strategies

for field traffic-induced subsoil compaction: a review: Part 1. Machine/soil interactions. Soil and tillage research, 73(1-2), 145-160.

87. Alakukku, L. (2000). Response of annual crops to subsoil compaction in a field experiment on clay soil lasting 17 years. Advances in GeoEcology, (32), 205-208.

88. Simojoki, A., Jaakkola, A., & Alakukku, L. (1991). Effect of compaction on soil air in a pot experiment and in the field. Soil and Tillage Research, 19(2-3), 175-186.

89. Ball, B. C., Scott, A., & Parker, J. P. (1999). Field N2O, CO2 and CH4 fluxes in relation to tillage, compaction and soil quality in Scotland. Soil and Tillage Research, 53(1), 29-39.

90. Fullen, M. A. (1985). Compaction, hydrological processes and soil erosion on loamy sands in east Shropshire, England. Soil and tillage research, 6(1), 17-29.

91. Hansen, S. (1996). Effects of manure treatment and soil compaction on plant production of a dairy farm system converting to organic farming practice. Agriculture, ecosystems & environment, 56(3), 173-186.

92. Radford, B. J., Yule, D. F., McGarry, D., & Playford, C. (2001). Crop responses to applied soil compaction and to compaction repair treatments. Soil and Tillage Research, 61(3-4), 157-166.

93. Fageria, N. K., Baligar, V. C., & Clark, R. (2006). Physiology of crop production. crc Press.

94. Masle, J. (2002). High soil strength: mechanical forces at play on root morphogenesis and in root: shoot signaling. Plant roots: the hidden half, (Ed. 3), 807-819.

95. Grzesiak, S., Grzesiak, M. T., Filek, W., Hura, T., & Stabryła, J. (2002). The impact of different soil moisture and soil compaction on the growth of triticale root system. Acta Physiologiae Plantarum, 24(3), 331-342.

96. Iijima, M., Kono, Y., Yamauchi, A., & Pardales Jr, J. R. (1991). Effects of soil compaction on the development of rice and maize root systems. Environmental and Experimental Botany, 31(3), 333-342.

97. Ojha, T. P., & Michael, A. M. (2006). Principles of agricultural engineering. Jain Brothers, New Delhi India, Vol. 1.

98. Murty, V. V. N. (1985). Land water management engineering.

99. Nieuwenhuis, H. S., & Schokking, F. (1997). Land subsidence in drained peat areas of the Province of Friesland, The Netherlands.

Quarterly Journal of Engineering Geology and Hydrogeology, 30(1), 37-48.

100. Schothorst, C. J. (1982). Drainage and behaviour of peat soils (No. 3). Wageningen, The Netherlands: ICW.

101. Ingebritsen, S. E., McVoy, C., Glaz, B., & Park, W. (1999). Florida Everglades. Land Subsidence in the United States, D. Galloway, DR Jones, and SE Ingebritsen, eds. USGS Circular, 1182, 96-106.

102. Rojstaczer, S., & Deverel, S. J. (1995). Land subsidence in drained histosols and highly organic mineral soils of California. Soil Science Society of America Journal, 59(4), 1162-1167.

103. Wösten, J. H. M., Ismail, A. B., & Van Wijk, A. L. M. (1997). Peat subsidence and its practical implications: a case study in Malaysia. Geoderma, 78(1-2), 25-36.

104. Stephens JC, Allen JH Jr, Chen E (1984) Organic soil subsidence. Geol Soc Am Rev Eng Geol 6:107–122

105. Deverel, S. J., & Rojstaczer, S. (1996). Subsidence of agricultural lands in the Sacramento-San Joaquin Delta, California: Role of aqueous and gaseous carbon fluxes. Water Resources Research, 32(8), 2359-2367.

106. Segeren, W. A., & Smits, H. (1974). Drainage of newly reclaimed marine clayey sediments, peat so ils, and acid-sulphate soils. Publication.

107. Dregne, H. E. (1977). Desertification of arid lands. Economic Geography, 53(4), 322-331.

108. Perini, L., Ceccarelli, T., Zitti, M., & Salvati, L. (2009). Insight desertification process: bio-physical and socio-economic drivers in Italy. Ital. J. Agrometeorol, 3, 45-55.

109. Bied-Charreton, M. (2008). Integrating the combat against desertification and land degradation into negotiations on climate change: a winning strategy. French Scientific Committee on Desertification.

110. UNCCD (United Nations Convention to Combat Desertification) (1994). United Nations convention to combat desertification in those countries experiencing serious drought and/or desertification particularly in Africa: text with annexes. UNEP, Nairobi

111. Hill, J., & Peter, D. (Eds.). (1996). The Use of Remote Sensing for Land Degradation and Desertification Monitoring in the Mediterranean Basin: State of the Art and Future Research: Proceedings of an Experts Workshop Jointly Organized by JRC/IRSA and DG XIII/D. 2, D. 4,

Valencia, Spain from 13 to 15 June 1994. European Commission, Directorate-General, Science, Research and Development.

112. Graetz, R. D. (1996). Empirical and practical approaches to land surface characterisation and change detection. The use of remote sensing for land degradation and desertification monitoring in the Mediterranean Basin. European Commission, Brussels, 9-22.
113. Culet P (2002) Desertification. In: Encyclopedia of life support systems. EOLSS Publishers Co. Ltd., Oxford
114. Keogh, M. E., & Törnqvist, T. E. (2019). Measuring rates of present-day relative sea-level rise in low-elevation coastal zones: a critical evaluation. Ocean Science, 15(1), 61-73.
115. Johnson, J. M., Moore, L. J., Ells, K., Murray, A. B., Adams, P. N., MacKenzie III, R. A., & Jaeger, J. M. (2015). Recent shifts in coastline change and shoreline stabilization linked to storm climate change. Earth Surface Processes and Landforms, 40(5), 569-585.
116. Batir, J. F., Hornbach, M. J., & Blackwell, D. D. (2017). Ten years of measurements and modeling of soil temperature changes and their effects on permafrost in Northwestern Alaska. Global and Planetary Change, 148, 55-71.
117. Taufik, M., Torfs, P. J., Uijlenhoet, R., Jones, P. D., Murdiyarso, D., & Van Lanen, H. A. (2017). Amplification of wildfire area burnt by hydrological drought in the humid tropics. Nature Climate Change, 7(6), 428-431.
118. Hellmann, J. J., Byers, J. E., Bierwagen, B. G., & Dukes, J. S. (2008). Five potential consequences of climate change for invasive species. Conservation biology, 22(3), 534-543.
119. Olsson, L., Barbosa, H., Bhadwal, S., Cowie, A., Delusca, K., Flores-Renteria, D., ... & Stringer, L. (2019). Land degradation. Climate Change and Land: an IPCC special report on climate change, desertification, land degradation, sustainable land management, food security, and greenhouse gas fluxes in terrestrial ecosystems, 345-436.
120. Pimm, S. L., Lawton, J. H., & Cohen, J. E. (1991). Food web patterns and their consequences. Nature, 350(6320), 669-674.
121. Hunt, H. W., Coleman, D. C., Ingham, E. R., Ingham, R. E., Elliott, E. T., Moore, J. C., ... & Morley, C. R. (1987). The detrital food web in a shortgrass prairie. Biology and Fertility of Soils, 3(1), 57-68.
122. Mikola, J., & Setälä, H. (1998). No evidence of trophic cascades in an

experimental microbial-based soil food web. Ecology, 79(1), 153-164.

123. David Hopkins, C. (2002). Managing wet soils: mole drainage.

124. Dickey, E. C., Shelton, D. P., & Jasa, P. J. (1986). Residue management for soil erosion control. Neb Guide G81-544-A. Inst. Agric. and Nat. Res., Univ. of Nebraska, Lincoln. Lincoln, NE.

125. Wischmeier, W. H., & Smith, D. D. (1978). Predicting rainfall erosion losses: a guide to conservation planning (No. 537). Department of Agriculture, Science and Education Administration.

126. Myers, R. J. K., & De Pauw, E. (1995). Strategies for the management of soil acidity. In Plant-Soil Interactions at Low pH: Principles and Management (pp. 729-741). Springer, Dordrecht.

127. Shaw, R.J. Hughes, K.K., Thorburn, P.J. & Dowling, A.J. (1987). 'Principles of landscape, soil and water salinity—processes and management options', Part A in Landscape, soil and water salinity, Proceedings of the Brisbane Regional Salinity Workshop, Brisbane, May 1987, Conference and Workshop Series QC87003, Department of Primary Industries, Queensland.

128. Johnston, A. M., & Bruulsema, T. W. (2014). 4R nutrient stewardship for improved nutrient use efficiency. Procedia Engineering, 83, 365-370.

129. Bufon, V. B. (2010). Optimizing Subsurface Drip Irrigation Design and Management with Hydrus-2D/3D Model. Texas Tech University, Vinicius Bof Bufon.

130. Ferguson, R. B., & Hergert, G. W. (2000). EC00-154 Precision Agriculture: Soil Sampling for Precision Agriculture. Historical Materials from University of Nebraska-Lincoln Extension, 708.

131. Smith, M. E., Facelli, J. M., & Cavagnaro, T. R. (2018). Interactions between soil properties, soil microbes and plants in remnant-grassland and old-field areas: a reciprocal transplant approach. Plant and Soil, 433(1), 127-145.

INDEX

D

E

F

G

Printed in the United States
by Baker & Taylor Publisher Services